THE CONCISE HANDBOOK OF

ANALYTICAL SPECTROSCOPY

Theory, Applications, and Reference Materials

Volume 1

Ultraviolet Spectroscopy

THE CONCISE HANDBOOK OF ANALYTICAL SPECTROSCOPY

Theory, Applications, and Reference Materials

Volume 1

Ultraviolet Spectroscopy

Jerry Workman

Unity Scientific, USA and National University, USA

World Scientific

NEW JERSEY • LONDON • SINGAPORE • BEIJING • SHANGHAI • HONG KONG • TAIPEI • CHENNAI • TOKYO

Published by

World Scientific Publishing Co. Pte. Ltd.

5 Toh Tuck Link, Singapore 596224

USA office: 27 Warren Street, Suite 401-402, Hackensack, NJ 07601

UK office: 57 Shelton Street, Covent Garden, London WC2H 9HE

Library of Congress Cataloging-in-Publication Data

Workman, Jerry.

The concise handbook of analytical spectroscopy : theory, applications, and reference materials / by Jerry Workman (Unity Scientific, USA & National University, USA).

volumes cm

Includes bibliographical references and index.

ISBN 978-9814508056 (Set)

ISBN 978-9814508087 (v. 1 : hardcover : alk. paper)

ISBN 978-9814508094 (v. 2 : hardcover : alk. paper)

ISBN 978-9814508100 (v. 3 : hardcover : alk. paper)

ISBN 978-9814508117 (v. 4 : hardcover : alk. paper)

ISBN 978-9814508124 (v. 5 : hardcover : alk. paper)

1. Spectrum analysis. I. Title.

QD95.W67 2015

543'.5--dc23

2014042606

British Library Cataloguing-in-Publication Data

A catalogue record for this book is available from the British Library.

Typeset by Stallion Press

Email: enquiries@stallionpress.com

Printed in Singapore

Preface

The Concise Handbook of Analytical Spectroscopy: Theory, Applications, and Reference Materials (In 5 Volumes)

Overview and Description

The concept of improving the use of electromagnetic energy to achieve a variety of qualitative and quantitative spectroscopic measurements on solid and liquid materials has been proliferating at a rapid rate. The use of such technologies to measure chemical composition and appearance; for classification; and to achieve a detailed understanding of material interactions has prompted a dramatic expansion in the use and development of spectroscopic techniques over a variety of academic and commercial fields. This multi-volume handbook is designed specifically as a reference tool for students, commercial development and quality scientists, and researchers or technologists in a variety of measurement endeavors. Each volume covers the theory, instrumentation, sampling methods, experimental design, and data analysis techniques, as well as essential reference tables, figures, and spectra for each spectroscopic region. The detailed practical aspects of applying spectroscopic tools for many of the most exciting and current applications are covered. Featured applications are included as well as common commercial analysis methods (i.e., color, appearance), spectroscopic quantitative and qualitative techniques, and advanced methods.

Outline

Each volume covers a specific region of the electromagnetic spectrum used for electronic and vibrational spectroscopic measurements. The volumes are not a series of essays, or a compilation of isolated papers on specific topics, but represent an integrated whole. The Handbook is compiled as a highly useful reference set for those directly involved in making and understanding spectroscopic measurement techniques. Each volume contains a logical progression of the material as follows.

Organization of Contents for Each Volume

Each volume commences with Front Reference Material for General Spectroscopy, preceding the following chapters.

1. General Introduction and History of Technique (includes useful basic concepts)
2. Theory of Technique (including delineation of common equations and references)
3. Types of Instrumentation (Types and design aspects)
4. Sampling Considerations (Cookbook approach to preparing samples)
5. Spectrum Data Acquisition (Experimental design and methods for data acquisition)
6. Data Analysis and Presentation Methods (The chemometrics and data analysis techniques applicable for each spectral region)
7. Applications and Spectral Examples (Cookbook approaches, results, and extensive literature reviews)
8. Reference Materials, Tables, Figures, and Spectra (The key look-up aspects for each technique)
9. Glossary of Specific Terms: Detailed Index for Reference Material and Text (Each volume includes a detailed table of contents and index that delineates the minutia of using the text and reference material in each volume.)

Each volume contains specifics for the spectral region covered, such as: common and unusual sampling devices; sampling methods (i.e., reagents, solvents, and sampling cells); and reference standard materials. Optics include: filters, etalons, prisms, and lenses; and fiber optics (principles, equipment, and connectors). In addition, reference tables and spectra, glossaries of terms, common equations and calculation methods, cookbook directions, and solved problems are included.

Reference Materials

Fundamental reference sources and early publications are referred to whenever possible as a primary source of data and information. Reference material common to each method is summarized within each volume to provide a more practical set of volumes for bench use. The intention is that each volume will stand somewhat alone for the working analyst using a specific spectroscopic method, such that required referral to other volumes is kept to a minimum and that key information is available regarding the individual spectroscopic technique within each volume.

Terminology and Symbols

Terminology use is included within the chapters and glossary sections for each volume. Note that *spectrophotometer* is the precise term used for each instrument described within the five volumes. However for brevity, the term *spectrometer* is sometimes used less precisely. Also, the symbols used are from different disciplines and thus there are some symbols used for more than one variable or parameter, depending on the common usage and the specific academic discipline where the symbology was formalized. A list of symbols used in this set of volumes may be located using the index and front material of each volume. The terms used for each equation within the volumes are defined where the equation is used.

Dedication

This book is dedicated to my fabulous wife Rebecca and to my family who have patiently sacrificed many hours of their lives for the cause of spectroscopy; and to everyone who spends many hours in laboratories analyzing samples for new knowledge, applied research, quality improvement, and the overall betterment of the lives of people … And to the One Great Designer who made all this marvelous measurement with light possible, discoverable, and such a pleasure to work with …

Author Biography

Jerome (Jerry) J. Workman, Jr. is Executive Vice President of Research & Engineering for Unity Scientific; Certified Core Adjunct Professor at National University, CA; and Principal at Biotechnology Business Associates. He was formerly Vice President of Technology Research for Masimo Corporation; Director of Research, Technology & Applications Development for Molecular Spectroscopy & Microanalysis for ThermoFisher Scientific; Senior Research Fellow at Kimberly-Clark Analytical Science & Technology; and Principal Scientist at Perkin-Elmer.

Dr. Workman has played a major role in defining and developing over twenty scientific instrument advancements and applications of novel vibrational spectroscopy technology for laboratory, process and remote monitoring of commercial synthetic and natural products and processes; and medical applications of spectroscopy. Over 55 U.S. and International Patent Applications (since 1998); 17 U.S. and International Patents issued, and multiple trade secrets. He has a total of 475 Technical Publications; and 18 reference books on a broad range of spectroscopy, chemometrics, and data processing techniques. He has received awards from the Eastern Analytical Symposium, ASTM International, and the Coblentz Society, as well as multiple fellowships, technical, and government appointments. He has taught annual courses in spectroscopy, chemometrics, and statistics for the Association of Official Analytical Chemists, the American Chemical Society, the Instrument Society of America, and the Federation of Analytical Chemists and Spectroscopy Societies; and at several universities, government agencies, and corporations.

Patented inventions include: multi-spectral monitoring (data fusion), web-based information technology in spectroscopy, scanning systems for positioning materials in manufacturing technology, invention of small molecule metabolite reporters for monitoring multiple parameters in human metabolism, advanced technology and algorithm development for instrument automation, standardization, and calibration, and for manufacturing monitoring methods for consumer products.

He is a Fellow of the American Institute of Chemists, the American Society for Testing and Materials International, and the Royal Society of Chemistry (U.K.). He is also a Chartered Chemist (CChem) of the Royal Society of Chemistry (England) and a Chartered Scientist of the Science Council (UK). He has served as Chair of the Industrial Advisory Board for the Center for Process Analytical Chemistry (CPAC) at the Univ. of Washington; and The Council for Near-Infrared Spectroscopy; and is past Chairman of ASTM Main Committee E13 on Molecular Spectroscopy and Separation Science. He has served on the U.S. National Academies NRC Panel for assessment of NIST programs.

With a B.A. in natural sciences and an M.A. in biological sciences from Saint Mary's University of Minnesota, Dr. Workman's Ph.D. is in biological chemistry from Columbia Pacific University. He is a graduate of the Columbia Senior Executive Program and also holds Columbia Business School Certificates in Executive Development (CIED) and in Business Excellence (CIBE). He also holds a Certificate in Strategy and Innovation from the M.I.T. Sloan School. He is listed in Who's Who in the World, Who's Who in America, and Who's Who in Science and Engineering.

Contents

Reference Tables: Comparison of Spectroscopic Techniques, the Electromagnetic Spectrum, Raman Analysis, and Useful Spectral Ranges for Sources, Detectors and Optical Materials

0.1 Physical Constants and Conversion Factors for SI and non-SI Units

Table 0.1 Spectroscopy constants.[1]

Physical constant	Value	Standard uncertainty	Concise form
Avogadro constant. N_A, L	$6.022\ 141\ 29\times10^{23}$ mol^{-1}	$0.000\ 000\ 27\times10^{23}$ mol^{-1}	$6.022\ 141\ 29(27)\times10^{23}$ mol^{-1}
Boltzmann constant, *k*	$1.380\ 6488\times10^{-23}$ J K^{-1}	$0.000\ 0013\times10^{-23}$ J K^{-1}	$1.380\ 6488(13)\times10^{-23}$ J K^{-1}
Faraday constant, *F*	96 485.3365 C mol^{-1}	0.0021 C mol^{-1}	96 485.3365(21) C mol^{-1}
Molar gas constant, *R*	8.314 4621 J mol^{-1} K^{-1}	0.000 0075 J mol^{-1} K^{-1}	8.314 4621(75) J mol^{-1} K^{-1}
Planck constant, *h*	$6.626\ 069\ 57\times10^{-34}$ J s	$0.000\ 000\ 29\times10^{-34}$ J s	$6.626\ 069\ 57(29)\times10^{-34}$ J s
Speed of light (in vacuum), *c*	299 792 458 m s^{-1}	(exact)	299 792 458 m s^{-1}
Stefan-Boltzmann constant, ö	$5.670\ 373\times10^{-8}$ W m^{-2} K^{-4}	$0.000\ 021\times10^{-8}$ W m^{-2} K^{-4}	$5.670\ 373(21)\times10^{-8}$ W m^{-2} K^{-4}

Table 0.2 Wavelength conversions.[2]

To convert Column 1 into Column 2 multiply by	Column 1 SI Units	Column 2 non-SI Units	To convert Column 2 into Column 1 multiply by
2.54	Centimeter, cm (10^{-2} m)	Inch, in	0.394
25.4	Millimeter, mm (10^{-3} m)	Inch, in	3.94×10^{-2}
1.0	Micrometer, μm (10^{-6} m)	Micron, μ	1.0
0.1	Nanometer, nm (10^{-9} m)	Angstrom, A (10^{-10} m)	10
$1/(\text{nm} \times 10^7)$	Nanometer, nm (10^{-9} m)	Wavenumber (cm^{-1})	$1/(\text{cm}^{-1} \times 10^7)$
$1/(\text{nm} \times 2.99792458 \times 10^{17})$	Nanometer, nm (10^{-9} m)	Frequency (Hertz, Hz)	$1/(\text{Hz} \times 2.99792458 \times 10^{17})$

Example 1: To convert Nanometer, nm (10^{-9} m) in Column 1 to Angstrom, A (10^{-10} m) in Column 2 to multiply by 0.1; therefore 0.1 nm = 1 Angstrom.

Example 2: To convert Angstrom, A (10^{-10} m) in Column 2 to Nanometer, nm (10^{-9} m) in Column 1 multiply by 10; therefore 10 Angstroms = 1 nm.

Table 0.3 Length conversions.

To convert Column 1 into Column 2 multiply by	Column 1 SI Units	Column 2 non-SI Units	To convert Column 2 into Column 1 multiply by
1.609	Kilometer, km (10^3 m)	Mile, mi	0.6213
0.914	Meter, m	Yard, yd	1.094
0.304	Meter, m	Foot, ft	3.28
1.0	Micrometer, μm (10^{-6} m)	Micron, μ	1.0
1000	Micrometer, μm (10^{-6} m)	Millimeter, mm (10^{-3} m)	0.001
25.4	Millimeter, mm (10^{-3} m)	Inch, in	3.94×10^{-2}
0.1	Nanometer, nm (10^{-9} m)	Angstrom, A	10

Table 0.4 Area conversions.

To convert Column 1 into Column 2 multiply by	Column 1 SI Units	Column 2 non-SI Units	To convert Column 2 into Column 1 multiply by
0.405	Hectare, ha	Acre	2.47
4.05×10^{-3}	Square kilometer, km^2 $(10^3\ m)^2$	Acre	2.47
2.590	Square kilometer, km^2 $(10^3\ m)^2$	Square mile, mi^2	0.386
4.05×10^{3}	Square meter, m^2	Acre	2.47×10^{-4}
9.29×10^{-2}	Square meter, m^2	Square foot, ft^2	10.76
645	Square millimeter, mm^2 $(10^{-3}\ m)^2$	Square inch, in^2	1.55×10^{-3}

Table 0.5 Volume conversions

To convert Column 1 into Column 2 multiply by	Column 1 SI Units	Column 2 non-SI Units	To convert Column 2 into Column 1 multiply by
102.8	Cubic meter, m^3	Acre-inch	9.73×10^{-3}
2.83×10^{-2}	Cubic meter, m^3	Cubic foot. ft^3	35.3
1.64×10^{-5}	Cubic meter, m^3	Cubic inch, in^3	6.10×10^{4}
35.24	Liter, L $(10^{-3}\ m^3)$	Bushel, bu	2.84×10^{-2}
0.946	Liter, L $(10^{-3}\ m^3)$	Quart (liquid), qt	1.057
28.3	Liter, L $(10^{-3}\ m^3)$	Cubic foot, ft^3	3.53×10^{-2}
3.78	Liter, L $(10^{-3}\ m^3)$	Gallon	0.265
2.96×10^{-2}	Liter, L $(10^{-3}\ m^3)$	Ounce (fluid), oz	33.78
0.473	Liter, L $(10^{-3}\ m^3)$	Pint (fluid), pt	2.11

Table 0.6 Mass conversions.

To convert Column 1 into Column 2 multiply by	Column 1 SI Units	Column 2 non-SI Units	To convert Column 2 into Column 1 multiply by
453.6	Gram, g (10^{-3} kg)	Pound, lb	2.2×10^{-3}
28.4	Gram, g (10^{-3} kg)	Ounce (Avoirdupois, avdp), oz	3.52×10^{-2}
0.454	Kilogram, kg	Pound, lb	2.205
100	Kilogram, kg	Quintal (metric), q	0.01
907	Kilogram, kg	Ton (2000 lb), U.S. ton	1.10×10^{-3}
0.907	Megagram, Mg (Tonne)	Ton (U.S.), ton	1.102
0.907	Tonne, t	Ton (U.S.), ton	1.102

Table 0.7 Rate conversions.

To convert Column 1 into Column 2 multiply by	Column 1 SI Units	Column 2 non-SI Units	To convert Column 2 into Column 1 multiply by
0.447	Meter per second, m s^{-1}	Mile per hour	2.237

Table 0.8 Specific surface conversions.

To convert Column 1 into Column 2 multiply by	Column 1 SI Units	Column 2 non-SI Units	To convert Column 2 into Column 1 multiply by
0.1	Square meter per kilogram, $m^2\,kg^{-1}$	Square centimeter per gram, $cm^2\,g^{-1}$	10
0.001	Square meter per kilogram, $m^2\,kg^{-1}$	Square millimeter per gram, $mm^2\,g^{-1}$	1000

Table 0.9 Density conversions.

To convert Column 1 into Column 2 multiply by	Column 1 SI Units	Column 2 non-SI Units	To convert Column 2 into Column 1 multiply by
1.00	Megagram per cubic meter, Mg m^{-3}	Gram per cubic centimeter, g cm^{-3}	1.00

Table 0.10 Pressure conversions.

To convert Column 1 into Column 2 multiply by	Column 1 SI Units	Column 2 non-SI Units	To convert Column 2 into Column 1 multiply by
0.101	Megapascal, MPa (10^6 Pa)	Atmosphere	9.869
0.1	Megapascal, MPa (10^6 Pa)	Bar	10
47.9	Pascal, Pa	Pounds per square foot, lb ft^{-2}	2.09×10^{-2}
6.90×10^3	Pascal, Pa	Pounds per square inch, lb in^{-2}	1.45×10^{-4}

Table 0.11 Temperature conversions.

To convert Column 1 into Column 2 multiply by	Column 1 SI Units	Column 2 non-SI Units	To convert Column 2 into Column 1 multiply by
100(K-273)	Kelvin, K	Celsius, °C	1.00 (°C + 273)
(°R − 491.67) × 5/9	Rankine, °R	Celsius, °C	(°C + 273.15) × 9/5
(9/5 °C) + 32	Celsius, °C	Fahrenheit, °F	5/9 (°F − 32)
°R − 459.67	Rankine, °R	Fahrenheit, °F	°F + 459.67
[K] × 9/5 − 459.67	Kelvin, K	Fahrenheit, °F	(°F + 459.67) × 5/9
[K] × 9/5	Kelvin, K	Rankine, °R	[°R] × 5/9

Table 0.12 Energy, work, quantity of heat conversions.

To convert Column 1 into Column 2 multiply by	Column 1 SI Units	Column 2 non-SI Units	To convert Column 2 into Column 1 multiply by
1.05×10^3	Joule, J	British thermal unit, Btu	9.48×10^{-4}
4.19	Joule, J	Calorie, cal	0.239
10^{-7}	Joule, J	erg	107
1.36	Joule, J	Foot-pound	0.735
4.19×10^4	Joule per square meter, $J\,m^{-2}$	Calorie per square centimeter (Langley)	2.387×10^{-5}
10^{-5}	Newton, N	Dyne	10^5
698	Watt per square meter, $W\,m^{-2}$	Calorie per square centimeter minute (irradiance), cal $cm^{-2}\,m^{-1}$	1.43×10^{-3}

Table 0.13 Plane angle conversions.

To convert Column 1 into Column 2 multiply by	Column 1 SI Units	Column 2 non-SI Units	To convert Column 2 into Column 1 multiply by
$1.75 \times 10^{-2} = \pi/180$	Radian, rad	Degrees (angle), °	57.3
60	Minutes (of arc), arcminute	Degrees (angle), °	$1/60 = 0.01667$
3600	Seconds (of arc), arcsecond	Degrees (angle), °	$1/3600 = 0.0002778$

Table 0.14 Electricity and magnetism conversions.

To convert Column 1 into Column 2 multiply by	Column 1 SI Units	Column 2 non-SI Units	To convert Column 2 into Column 1 multiply by
10	Siemen per meter, S m-1	Millimho per centimeter, mmho cm^{-1}	0.1
10^{-4}	Tesla, T	Gauss, G	10^4

Table 0.15 Fluid flow conversions.

To convert Column 1 into Column 2 multiply by	Column 1 SI Units	Column 2 non-SI Units	To convert Column 2 into Column 1 multiply by
102.8	Cubic meter, m^3	Acre-inch, acre-in	9.73×10^{-3}
101.9	Cubic meter per hour, $m^3 h^{-1}$	Cubic foot per second, $ft^3 s^{-1}$	9.81×10^{-3}
0.227	Cubic meter per hour, $m^3 h^{-1}$	U.S. gallon per minute, gal min^{-1}	4.40
0.123	Hectare meter, ha m	Acre-foot, acre-ft	8.11
1.03×10^{-2}	Hectare meter, ha m	Acre-inch, acre-in	97.28
12.33	Hectare centimeter, ha cm	Acre-foot, acre-ft	8.1×10^{-2}

Table 0.16 Concentration conversions.

To convert Column 1 into Column 2 multiply by	Column 1 SI Units	Column 2 non-SI Units	To convert Column 2 into Column 1 multiply by
1	Centimole per kilogram, cmol kg^{-1}	Milliequivalent per 100 grams, meq 100 g^{-1}	1
0.1	Gram per kilogram, g, kg^{-1}	Percent, %	10
1	Milligrams per kilogram, mg kg^{-1}	Parts per million, ppm	1

Table 0.17 Radioactivity conversions.

To convert Column 1 into Column 2 multiply by	Column 1 SI Units	Column 2 non-SI Units	To convert Column 2 into Column 1 multiply by
3.7×10^{10}	Becquerel, Bq	Curie, Ci	2.7×10^{-11}
37	Becquerel per kilogram, Bq kg^{-1}	Picocurie per gram, pCi g^{-1}	2.7×10^{-2}
0.01	Gray, Gy (Absorbed dose)	Rad, rd	100
0.01	Sievert, Sv (equivalent dose)	Rem (Roentgen equivalent man)	100

0.2 Introduction and Comparison of Spectroscopic Methods[3]

0.2.1 *Ultraviolet (UV) Spectroscopy*

The Ultraviolet region is specified as 190 to 360 nanometers (nm) or 10^{-9} meters. The types of electrons that can be excited by UV/Vis light are few in number: nonbonding electrons, electrons in single bonds, and electrons involved in double bonds. These may be excited to several excited states. The distinction between molecules is that the ability to "jump" to higher states is affected by attached moieties — for example: double bonds, conjugations, and elements such as: oxygen, bromine, and others with pairs of nonbonding electrons. As a consequence, most elements with UV/Vis absorbances have specific wavelengths at which their peak absorbances occur. These peaks may be used to identify a particular molecule. The UV/Vis region is not as "rich" in information as, say, the infrared spectrum, but it reveals enough detail to enable a comparison of a material with a previously identified substance. A common use of this capability is in the Pharmaceutical industry, where UV/Vis detectors are frequently used with HPLC instruments as a final check before a drug product is released for sale to consumers. The various moieties or chromophores associated with ultraviolet absorption include: Nitriles (R-C≡N), 160 nm; Acetylenes (-C≡C-), 170 nm; Alkenes (>C=C<), 175 nm; Alcohols (R-OH), 180 nm and 175–200 nm; Ethers (R-O-R′), 180 nm; Ketones (R-C=O -R′), 180 nm and 280 nm; Amines, primary ($R\text{-}NH_2$), 190 nm and 200–220 nm; Aldehydes (R-C=O-H), 190 nm and 290 nm; Carboxylic acids (R-C=O-OH), 205 nm; Esters (R-C=O-OR′), 205 nm; Amides, primary ($R\text{-}C{=}O\text{-}NH_2$), 210 nm; Thiols (R-SH), 210 nm; Nitrites ($R\text{-}NO_2$), 271 nm; and Azo-group (R-N=N-R′), 340 nm.

0.2.2 *Visible (Vis) Spectroscopy*

Visible light (360 to 780 nm) is the light or electromagnetic energy seen by the human eye. The actual light demonstrated by different colors is visible light and spectra in the visible region are subjected to specific mathematical algorithms to determine color coordinates or color space coordinates to enable specific colors and brightness or darkness to be precisely described in mathematical terms. The aspects of color measurement and computation are contained in the text volume on visible spectroscopy and may be located in the index for this set of volumes. The terms "color measurement" and "visible spectroscopy" are the main index topic headings to find detailed information in this set of volumes. Colors are produced by electrons in a pigment moving from one orbital transition to another around the atoms within the molecules of the colored substance. Visible reflectance or transmittance spectra demonstrate the various colors. The color appearance of the human eye to a particular visible spectrum is dependent on: the light source color temperature and emission spectrum, the observer angle of observation, and the object background brightness and color differences. The size, direction, and scattering/absorption characteristics of the object also have an effect on its color appearance. The basic seven colors first described by Isaac Newton, with their corresponding wavelength scale in nanometers, consist of: violet (360–415 nm); indigo (415–444 nm), blue (444–487 nm), green (487–540 nm), yellow (540–590 nm), orange (590–690 nm), and red (690–830 nm). A twelve color system, with corresponding wavelength information is more common in modern terminology, although technically color is defined by a series of color measurement coordinate systems. These color measurement coordinate systems include: XYZ tristimulus values, Yxy color space, L*a*b* color space, L*C*h color space, and Hunter Lab color space. These are described in detail within sections of these volumes (see index and coordinate system names for more information).

There are also molecular absorptions that occur within the visible region and these are described in various chapters within this text and include the notations for the harmonic band or overtone involved: O-H Alkyl alcohol (6ν, no hydrogen bonding), Aromatic C-H Stretch (5ν), O-H Alkyl alcohol (4ν, no hydrogen bonding), Methyl C-H Stretch (5ν), O-H Phenols (4ν, no hydrogen bonding), Methylene C-H Stretch (5ν), O-H Primary Alcohols (4ν), O-H Water (4ν), O-H Secondary Alcohols (4ν), O-H Tertiary Alcohols (4ν), and Alkenes, conjugated RC=C-C=C-R′. The visible spectra include information related to both electronic and molecular interactions of energy with atoms and molecules. Volume 2, Chapter 7 gives more details.

0.2.3 *Near-Infrared (NIR) Spectroscopy*

Near-infrared spectroscopy is used where multicomponent molecular vibrational analysis is required in the presence of interfering substances. The near-infrared spectra consist of overtones and combination bands of the fundamental molecular absorptions found in the mid-infrared region. Near-infrared spectra consist of generally overlapping vibrational bands that are non-specific and poorly resolved. The use of chemometric mathematical data processing can be used to calibrate for qualitative or quantitative analysis despite these apparent spectroscopic limitations. Traditional near-infrared spectroscopy was used in agricultural product analysis for lignin polymers (2270 nm), paraffins and long alkane chain polymers (2310 nm), glucose based polymers such as cellulose (2336 nm), amino acid polymers as proteins (2180 nm), carbohydrates (2100 nm), and moisture (1440 and 1940 nm). The dominant near-infrared spectral features include the methyl C-H stretching vibrations, methylene C-H stretching vibrations, aromatic C-H stretching vibrations, O-H stretching vibrations, methoxy C-H stretching, and carbonyl associated C-H stretching. In addition, N-H from primary amides, secondary amides (both alkyl, and aryl group associations), N-H from primary, secondary, and tertiary amines, and N-H from amine salts predominate near-infrared spectral features of polymers and organic compounds.

0.2.4 *Infrared (IR or Mid-IR) Spectroscopy*

Mid-infrared spectroscopy provides a measurement technique for intense, isolated and reliable absorption bands of fundamental molecular vibrations from polymers and other organic compounds. The spectrometric methodology allows for univariate calibration with higher signal strength (absorptivity) required for solid, liquids or gas phase measurements. Relatively small pathlengths of 0.1 to 1.0 mm are required for hydrocarbon liquids and solids. The technique is generally incompatible with the use of fiber optics, but specialized fiber materials exist. Instrumentation is higher cost than near-infrared spectrophotometers for the most part.

Dominant mid-infrared spectral include the C-H (methyl, methylene, aromatic, methoxy, and carbonyl) fundamental stretching and bending molecular vibrations, O-H (hydroxyl) stretch fundamental vibrations; N-H (amine) stretching, C-F (fluorocarbon) stretching, -C≡N (nitrile) stretching, -C=O (carbonyl) stretch from esters, acetates, and amides; C-Cl stretch from chlorinated hydrocarbons, and $-NO_2$ from nitro- containing compounds.

0.2.5 *Raman Spectroscopy*

Raman spectroscopy can be used for a variety of measurements on samples that are aqueous in nature or where glass sample holders are present. Carbon dioxide, water, and glass (silica) are weak scatterers and thus there is generally no problems in analyzing samples having these properties. There is typically no sample preparation involved in samples measured using Raman. Raman spectroscopy is complementary to mid infrared spectroscopy in the measurement of fundamental molecular vibrations. Raman measurements are compatible with fiber optics. Raman measurements exhibit high signal-to-noise and a reasonable cost for instrumentation. The dominant Raman spectral features are acetylenic C≡C stretching, olefinic C=C stretch at 1680–1630 cm^{-1}, N=N (azo-) stretching, S-H (thio-) stretching, C=S stretching, C-S stretching, and S-S stretching bands. Raman spectra also contain such molecular vibrational information as CH_2 twist and wagging, carbonyl

C=O stretch associated with esters, acetates, and amides; C-Cl (halogenated hydrocarbons) stretching, and -NO_2 (nitro-/nitrite) stretching. In addition, Raman yields information content of phenyl- containing compounds at $1000\,cm^{-1}$.

0.3 The Electromagnetic Spectrum

The electromagnetic spectrum consists of many different types of radiation (energy): gamma, x-ray, ultraviolet (UV), visible (Vis), infrared (IR), microwave, and radio waves (Figures 0.1 through 0.3). Each type of radiation occupies its own region of the electromagnetic spectrum, and the major difference between the individual spectral regions is merely the energy involved. This energy is expressed in units of: photon energy (eV), Frequency (Hz), wavenumber (cm^{-1}), or wavelength (nm) and the manner in which these energies interact with matter.

With the possible exception of gamma radiation, most people are familiar with these types of radiation through their experiences in daily living. X-rays penetrate our bodies, allowing physicians to visualize our internal anatomy. UV light is associated with sunburn and tanning. We see colors and objects in the visible spectrum. In the kitchen, we toast our bread with IR radiation, and "zap" our meals with microwaves. We use microwave transmission for cell phone technology and we use radio waves to broadcast sound and images through AM/FM radio and television signals. The spectral region most useful in day-to-day analytical chemistry is the range of wavelengths from just below our visual perception (i.e., Ultraviolet or UV) through the colors we see (i.e., visible or Vis), known collectively as the UV/Vis spectral region. The wavelengths covered in the UV/Vis are measured in nanometers (nm), a unit of length representing one billionth of a meter. The generally accepted ranges for the UV/Vis region are ultraviolet (190–380 nm) and visible (380–750 nm). Some UV/Vis work involves longer wavelengths in the neighboring shortwave near-IR spectral region, from about 750–950 nm.

0.3.1 *Molecular and Electronic Spectroscopy Unit Conversions*

The units used to describe spectra for molecular spectroscopy vary with the academic or engineering discipline describing the spectral regions. For the most part the spectra are described in terms of an optical response specific to energy from a particular spectral region. The responses are in terms of inelastic or elastic collisions of the energy with various materials and are referred to as scattering absorption, absorbance, reflection, transmission, and so on. The various spectral energy for molecular and electronic spectroscopy is described in terms of units of wavelength (e.g., microns (μm) as 10^{-6} meters, or nanometers (nm) as 10^{-9} m, or Angstroms (Å) as 10^{-10} m). Energy is also expressed as frequency (cycles per second or sec.$^{-1}$) termed Hertz (Hz), as Wavenumbers (cm^{-1}), and as electron Volts (eV).

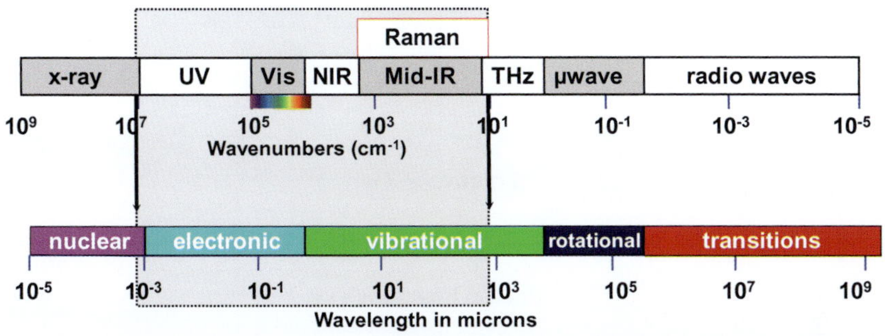

Fig. 0.1 The electromagnetic spectrum highlighting the ultraviolet to infrared/Raman spectral regions. Ultraviolet through Infrared/Raman range is shown in the rectangle. This region is from 190 nanometers to 25,000 nanometers, or 0.19 to 25 microns and represents electronic transitions through vibrational absorptions.

Molecular Electromagnetic Spectrum Regions

Region	Wavelength (nanometers)	Wavenumbers (cm^{-1})	Frequency (Hz)	Energy (eV)
Radio Waves	$1e^{9}$-$1e^{17}$	$1e^{-2}$-$1e^{-10}$	$3.0e^{8}$-3.0	$1.2e^{-6}$-$1.2e^{-14}$
Microwaves	$1e^{6}$-$1e^{9}$	10-0.01	$3.0e^{11}$-$3.0e^{8}$	$1.2e^{-3}$-$1.2e^{-6}$
Terahertz	2.5×10^{4}-10^{6}	400-10	$1.2e^{13}$-$3.0e^{11}$	0.05-0.001
Infrared	2500-25,000	4000-400	$1.2e^{14}$-$1.2e^{13}$	0.50-0.05
Raman	2500-25,000	4000-400	$1.2e^{14}$-$1.2e^{13}$	0.50-0.05
Near Infrared	780-2500	14,493-3,333	$3.8e^{14}$-$1.2e^{14}$	1.6-0.50
Visible	360-780	27,778-12,821	$8.3e^{14}$-$3.8e^{14}$	3.4-1.6
Ultraviolet	190-360	52,632-27,778	$1.6e^{15}$-$8.3e^{14}$	6.5-3.4
Vacuum UV	10-190	$1e^{6}$-$5.26e^{4}$	$3.0e^{16}$-$1.6e^{15}$	124-6.5
X-Rays	0.01-10	$1e^{9}$-$1e^{6}$	$3.0e^{19}$-$3.0e^{16}$	$124e^{3}$-124
Gamma Rays	<0.01	>$1e^{9}$	>$3e^{19}$	>$124e^{3}$

Fig. 0.2 Comparative Regions of the Electromagnetic Spectrum in wavelength (nanometers), wavenumbers, Frequency (Hertz), and Energy (electron Volts).

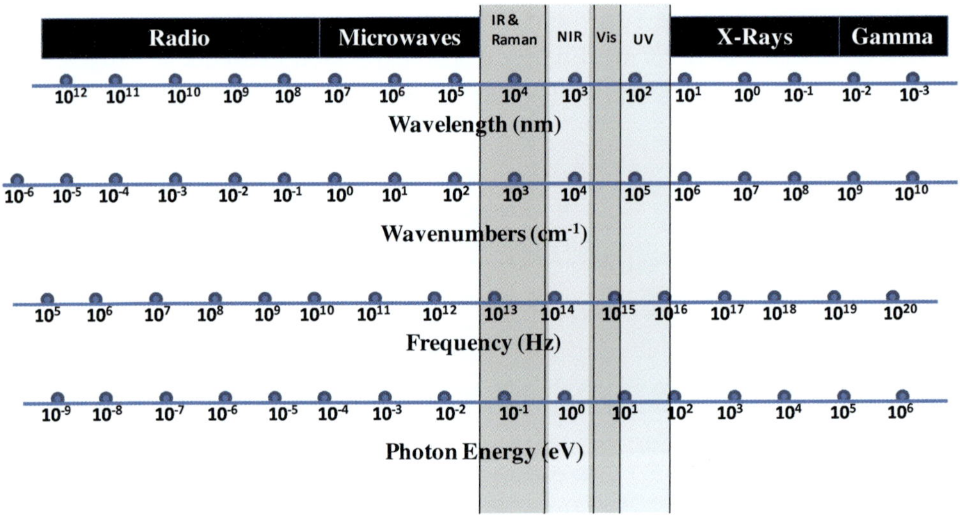

Fig. 0.3 Graphical Illustration of the Comparative Regions of the Electromagnetic Spectrum in wavelength (nanometers), wavenumbers, Frequency (Hertz), and Energy (electron Volts).

The various formulas used for the inter-conversion of units for spectroscopy include the following. Light has both particle and wave properties, quantum theory tells us that the energy of a light "particle" or photon E_p is given by:

$$E_p = h\nu \tag{0.1}$$

$$E_p = hc\bar{\nu} \tag{0.2}$$

where: h = Planck's constant (or 6.6256×10^{-27} erg-sec.); and ν (nu) is the frequency of light, also known as Hertz (Hz) or the number of vibrations per second, in units of sec.$^{-1}$; $\bar{\nu}$ is the wavenumber units as cm^{-1} (i.e., the number of waves per centimeter), c is the velocity of light in a vacuum (or 2.9979×10^{10} cm·sec^{-1}).

$$\bar{\nu} = \frac{\nu}{c \cdot n^{-1}} = \frac{1}{\lambda} \tag{0.3}$$

$$\nu = \frac{c \cdot n^{-1}}{\lambda} \tag{0.4}$$

where: $\bar{\nu}$ is the wavenumber units as cm^{-1} (i.e., the number of waves per centimeter), c is the velocity of light in a vacuum (or 2.9979×10^{10} cm·sec^{-1}), ν (Greek: nu) is the frequency of light, also known as Hertz (Hz) or the number of vibrations per second, in units of (sec.$^{-1}$), and n is the refractive index of the medium the light is passing through (e.g., air = 1.0003), λ (Greek: lambda) is the wavelength in units of cm. Combining the above equations we also note:

$$\bar{\nu} = \frac{hc}{E_p} \Rightarrow \frac{1}{\lambda} = \frac{hc}{E_p} \Rightarrow \lambda = \frac{E_p}{hc} \tag{0.5}$$

0.3.2 *Direct Unit Conversions*

Wavelength (in nanometers) to wavenumbers (cm^{-1}):

$$\bar{\nu}_{cm^{-1}} = \frac{10^7}{\lambda_{nm}} \tag{0.6}$$

Wavenumbers (cm^{-1}) to wavelength (in nanometers):

$$\lambda_{nm} = \frac{10^7}{\bar{\nu}_{cm^{-1}}} \tag{0.7}$$

Wavelength (in nanometers) to frequency (in Hz):

$$\nu_{HZ} = \frac{2.9979 \times 10^{17} nm \cdot \sec^{-1}}{\lambda_{nm}} \tag{0.8}$$

Wavelength (in nanometers) convert to energy (in eV):

$$E_{p_{eV}} = \frac{1240}{\lambda_{nm}} \tag{0.9}$$

0.4 Description of Basic Quantitative Spectroscopic Measurements using Beer's Law[4]

Spectroscopic measurements depend upon the principle that light energy interacting with a material will cause absorption at a specific frequency depending upon the chemical characteristics of that material. The amplitude of the absorption at any particular frequency (or wavelength or wavenumber) is determined by the absorptivity of the molecule being measured and the number of molecules encountered by the beam path of the measuring instrument. It is assumed that a change in spectral response is related to a concentration as described by the Bouguer, Lambert, and Beer relationship, most often termed Beer's law. The Beer's law relationship is described as the absorbance (A, Au, or signal strength) of an analyte being measured using a spectrophotometer is equivalent to the product of the absorptivity (ε) of a specific type of molecular vibration at a given frequency; the concentration (c) of the molecules in the measurement beam; and the pathlength (l) of the sample holder within the measurement beam. This relationship between measured spectral signal and concentration of a molecule is most often expressed as:

$$A = \varepsilon cl \tag{0.10}$$

where ε is the molar absorptivity (referred to as molar extinction coefficient by earlier physicists) in units of Liter·Mole^{-1}·cm^{-1}; c is the concentration of molecules in the spectrometer beam in units of Mole·Liter^{-1} (Note: this is a scaled volume fraction unit); and pathlength (l) is the thickness of the sample in units of cm of the measured sample at a specific concentration. The absorptivity for any specific molecule type is calculated by careful measurements of the absorbance of a compound, generally diluted in a suitable organic solvent, and by applying the relationship:

$$\varepsilon = \frac{A}{cl} \tag{0.11}$$

Note that for transmittance (where T = 0.0 to 1.0) and percent transmittance (where %T = 0 to 100.0) spectroscopy, a more complete delineation of the relationships between the various terms is contained in an expression such as:

$$T = \frac{I}{I_0} = 10^{-\varepsilon cl} \Rightarrow \text{Abs.} = A = -\log_{10}\left(\frac{I}{I_0}\right) = -\log_{10} T = \varepsilon cl \tag{0.12}$$

Here the symbols I and I_0 represent the attenuated energy detected after sample interaction and the initial energy incident to the sample, respectively. For reflectance (where %R = 0.0 to 1.0) and percent reflectance (Where R = 0.0 to 100.0) spectroscopy the various relationships are expressed as:

$$R = \frac{I}{I_0} = 10^{-\varepsilon cl} \Rightarrow Abs. = A = -\log_{10}\left(\frac{I}{I_0}\right) = -\log_{10} R = \varepsilon cl \tag{0.13}$$

Noting the relationship exists where the change in intensity (I) of the transmitted or reflected light from a sample is a function of the change in pathlength (l) of the sample as expressed by the absorptivity (ε) of a specific analyte (or molecular substance) and its concentration (c) by:

$$-\frac{\partial I}{\partial l} = \varepsilon cl(\ln 10) \tag{0.14}$$

Modern spectrophotometers utilize these assumptions for making spectroscopic measurement and generally display spectroscopic data as transmission (T), reflection (R), and absorbance A (y-axis or ordinate axis) versus wavelength (nm, microns) or wavenumber (cm^{-1}) (as x-axis, or abscissa axis).

0.5 Spectral Ranges for Sources, Detectors and Optical Materials

Table 0.18 Source emissivity ranges.[3]

Source emissivity ranges (μm)	Start	End
Quartz tungsten halogen	0.22	2.7
Glass tungsten halogen	0.25	2.25
DC deuterium lamp	0.185	3.75
Pulsed xenon arc lamp	0.18	2.5
DC arc lamp	0.20	2.5
Globar (Silicon Carbide)	1	100
Nernst glower	0.3	35
Carbon arc	0.5	100
Mercury lamp	0.3	100
Helium-neon laser (He:Ne)	0.6327	0.6328
Neodymium yttrium aluminum garnet (Nd:YAG) laser	1.0639	1.0640

Table 0.19 Detector sensitivity ranges.[3]

Detector sensitivity ranges (μm)	Start	End
Silicon (Si)	0.30	1.1
Lead Sulfide (PbS)	1.1	3
Indium Arsenide (InAs)	1.7	5.7
Indium Gallium Arsenide (InGaAs), typical	0.90	1.7
Indium Gallium Arsenide (InGaAs), extended	0.65	2.4
Indium Gallium Arsenide (InGaAs), IR extended	0.65	2.7
Germanium (Ge:X)	2	40
Germanium Gold (Ge:Au)	2	9
Germanium Cadmium (Ge:Cd)	2	24
Lead Selenide (PbSe)	1.7	5.5
Germanium Zinc (Ge:Zn)	2	40
Indium Antimonide (InSb)	1.8	6.8
Lead Telluride (PbTe)	1.5	4.5
Deuterated Triglycine Sulfate (DTGS/KBr)	0.83	25
Deuterated Triglycine Sulfate (DTGS/PE)	10	120
Mercury Cadmium Telluride (MCT)	1	17
Triglycine Sulfate (TGS)	10	120
Pyroelectric Lithium Tantalate ($LiTaO_3$) (PLT)	1.5	30
Silicon Photodiode Detectors	0.19	1.1
Photomultipler Tube (PMT), S1 Type Photocathode	0.3	1.2
Photomultipler Tube, Cesium Gallium Arsenide (CsGaAs) Photocathode	0.3	0.93
Photomultipler Tube, Cesium Indium Gallium Arsenide (CsInGaAs) Photocathode	0.3	1
Photomultipler Tube, Cesium Antiminide CsSb Photocathode	0.25	0.75
Photomultipler Tube, Multialkali (Na-K-Sb-Cs) Photocathode	0.3	0.93
Photomultipler Tube, CsI Photocathode	0.2	0.9
Photomultipler Tube, S1 Type Photocathode	0.3	1.2
Photomultipler Tube, Cesium Gallium Arsenide (CsGaAs) Photocathode	0.3	0.93
Photomultipler Tube, Cesium Indium Gallium Arsenide (CsInGaAs) Photocathode	0.3	1
Silicon (Si) CCD	0.2	1.1
Indium Gallium Arsenide (InGaAs) CCD	0.8	1.7
Ultraviolet range	0.19	0.36
Visible Range	0.36	0.78
Near-Infrared Range	0.78	2.5
Infrared	2.5	25
Raman	2.5	25

Table 0.20 Optical material spectral ranges.[3]

Optical material spectral ranges (μm)	Start	End
Methacrylate, Polymer	0.25	1.1
Silica, UV Grade Fused (SiO_2)	0.2	2.5
Silica, Synthetic Fused (SiO_2)	0.23	2.6
Quartz, Crystalline (SiO_2)	0.24	2.6
Quartz, extremely low O-H	0.19	2.7
Glass, Flint (SF 10)	0.38	2.35
Glass, Flint (SF 8)	0.355	2.35
Glass (BK 7)	0.315	2.35
Glass, "Optical Crown"	0.32	2.3
Glass, "Borosilicate Crown"	0.36	2.35
Glass, Pyrex®	0.31	2.5
Glass, Zerodur®Schott	0.4	2.85
Glass, Tempax®	0.36	2.35
Sapphire (Aluminum Oxide) (Al_2O_3) 5 mm is 50 %T at 5.1 microns	0.15	5.1
Sodium Chloride (NaCl)	0.25	16
Suprasil 300	0.19	3.6
Diamond (C)	0.22	4
Silica, Fused as pure SiO_2 (IR Quality)	0.2	4
Quartz, Anhydrous	0.2	2.75
Zirconium Fluoride (ZrF)	0.9	4.76
Chalcogenide (NIR-IR measurements)	2.22	11.1
Cesium Iodide (CsI)	0.3	50
Lead Sulfide (PbS)	1.1	3
Potassium Bromide (KBr)	0.25	26
Potassium Chloride (KCl)	0.25	20
Thallium Bromide-Iodide (KRS-5)	0.6	40
Germanium (Ge)	1.1	30
Arsenic Sulfide (As_2S_3)	0.6	15
Magnesium Fluoride (MgF_2)	0.6	9.5
Zinc Selenide (ZnSe) (IRTRAN-4)	0.6	26
Barium Fluoride (BaF_2)	0.2	13
Zinc Sulfide (ZnS) (Cleartran or IRTRAN-2)	0.6	15
Calcium Fluoride (CaF_2) (IRTRAN-3)	0.2	9
Silver Bromide (AgBr)	0.5	35
Polyethylene, Polymer (High Density)	16	333
Glass, Silica, Fused "Water Free" (SiO_2); 10 mm is 82%T at 3.15 microns	0.15	3.15
Glass, Corning Optical Glasses as HPFS®7978 and 7979 50%T/cm	0.15	4.35

(*Continued*)

Table 0.20 (*Continued*)

Optical material spectral ranges (μm)	Start	End
Glass, Corning Optical Glasses as HPFS® 7980; 50%T/cm	0.2	2.6
Glass, Corning Optical Glasses as HPFS® 865X and 797X 50%T/cm	0.15	2.6
Glass, ESCO Optical Glasses as S1-UVA and S1-UVB Working Range (Band 1)	0.17	2.2
Glass, ESCO Optical Glasses as S1-UVA and S1-UVB Working Range (Band 2)	2.9	3.6
Glass, ESCO Optical Glasses as I2-IR Working Range	0.25	3.6
Glass, ESCO Optical Glasses as G1 and A1 Working Range	0.27	2.5
Glass, ESCO Optical Glasses as BK7 Working Range	0.35	2
Ultraviolet range	0.19	0.36
Visible Range	0.36	0.78
Near-Infrared Range	0.78	2.5
Infrared	2.5	25
Raman	2.5	25

Table 0.21a UV-Vis gas emission sources and peak positions (in nanometers).[5]

Position	Source	Position	Source	Position	Source
253.652	Hg	452.186	Xe	612.884	Ne
296.728	Hg	462.42	Xe	614.306	Ne
302.15	Hg	466.849	Xe	616.359	Ne
313.155	Hg	469.097	Xe	621.728	Ne
334.148	Hg	469.804	Xe	626.649	Ne
341.79	Ne	473.415	Xe	630.479	Ne
342.391	Ne	479.262	Xe	633.442	Ne
344.77	Ne	480.702	Xe	638.299	Ne
345.076	Ne	482.971	Xe	640.225	Ne
345.419	Ne	484.329	Xe	650.653	Ne
346.052	Ne	491.651	Xe	653.288	Ne
346.658	Ne	492.315	Xe	659.895	Ne
347.257	Ne	503.135	Ne	667.828	Ne
349.806	Ne	503.775	Ne	671.704	Ne
350.121	Ne	508.038	Ne	692.947	Ne
351.519	Ne	511.367	Ne	703.241	Ne
352.047	Ne	511.65	Ne	717.394	Ne
359.353	Ne	540.056	Ne	724.512	Ne
360.017	Ne	546.074	Hg	733.93	Xe
363.366	Ne	556.222	Kr	738.6	Xe
365.015	Hg	557.029	Kr	739.379	Xe
368.573	Ne	576.441	Ne	740.04	Xe
370.122	Ne	576.96	Hg	743.89	Ne
404.656	Hg	579.066	Hg	747.244	Ne
407.783	Hg	582.015	Ne	748.887	Ne
431.958	Kr	585.249	Ne	753.577	Ne
435.833	Hg	587.096	Kr	754.404	Ne
436.264	Kr	588.189	Ne	755.979	Xe
437.612	Kr	594.483	Ne	758.468	Xe
439.997	Kr	597.553	Ne	758.741	Kr
445.392	Kr	602	Ne	760.155	Kr
446.369	Kr	607.433	Ne	764.391	Xe
450.235	Kr	609.616	Ne	768.525	Kr
				769.454	Kr

Table 0.21b NIR gas emission sources and peak positions (in nanometers).[5]

Position	Source	Position	Source	Position	Source
780.265	Xe	1107.887	Ar	1390.748	Ar
785.482	Kr	1117.752	Ne	1409.364	Ar
788.132	Xe	1144.183	Ar	1414.244	Xe
791.343	Kr	1148.811	Ar	1442.679	Kr
796.734	Xe	1152.275	Ne	1473.281	Xe
805.726	Xe	1166.871	Ar	1473.444	Kr
805.95	Kr	1171.949	Ar	1504.65	Ar
806.134	Xe	1181.938	Kr	1517.269	Ar
810.436	Kr	1211.233	Ar	1520.31	Kr
819.006	Kr	1213.974	Ar	1532.934	Ar
823.163	Xe	1220.353	Kr	1537.204	Kr
826.324	Kr	1234.339	Ar	1541.839	Xe
826.652	Xe	1243.932	Ar	1598.949	Ar
829.811	Kr	1248.766	Ar	1605.328	Xe
837.761	Ne	1262.339	Xe	1620.872	Kr
849.536	Ne	1270.228	Ar	1647.29	Xe
866.794	Ar	1273.342	Ar	1656.023	Xe
877.675	Kr	1280.274	Ar	1672.815	Xe
878.375	Ne	1295.666	Ar	1689.676	Kr
881.941	Xe	1300.826	Ar	1694.058	Ar
892.869	Kr	1317.741	Kr	1704.288	Ar
904.545	Xe	1322.811	Ar	1755.35	Kr
912.297	Ar	1322.811	Ar	1763.882	Xe
916.265	Xe	1327.264	Ar	1785.738	Kr
922.45	Ar	1331.321	Ar	1790.45	Xe
935.422	Ar	1336.711	Ar	1800.223	Kr
965.779	Ar	1350.419	Ar	1809.09	Xe
978.45	Ar	1362.266	Ar	1816.733	Kr
979.97	Xe	1363.422	Kr	1832.53	Xe
992.319	Xe	1365.706	Xe	1959.94	Xe
1047.005	Ar	1367.855	Ar	1984.638	Xe
1067.357	Ar	1371.858	Ar	2190.851	Kr
1083.837	Xe	1382.572	Ar		

References

1. NIST: http://physics.nist.gov/cuu/Constants/index.html.
2. Adapted from Near-Infrared Spectroscopy in Agriculture xx-xxiii (Used with Permission). Tables have been changed with different values and conversion factors.
3. J. Workman, *The Academic Press Handbook of Organic Compounds: NIR, IR, Raman, and UV-VIS Spectra Featuring Polymers, and Surfactants*, (3 Volume Set: Vol. 1, Methods and Interpretation; Vol. 2, UV-Vis and NIR Spectra; Vol. 3, IR and Raman Spectra), Academic Press, Boston (2000).
4. J. Workman and L. Weyer, *Practical Guide and Spectral Atlas to Interpretive Near-Infrared Spectroscopy*, 2nd edn (CRC — Taylor and Francis, Boca Raton, FL 2012).
5. Ocean Optics Wavelength Calibration Sources, www.oceanoptics.com. (Used with Permission).

Chapter 1

Introduction to Ultraviolet Spectroscopy

Contents

1.1 Introduction

As modern instrumentation for Ultraviolet-Visible (UV-Vis) spectroscopy becomes increasingly automated and easy to use, laboratories are now able to assign the routine analysis of samples to workers just getting started in their laboratory careers. This section is intended to describe the basic tenets of UV-Vis spectroscopy, establishing a framework for further learning by providing a basic introduction. This includes an understanding of the electromagnetic spectrum and how radiation interacts with matter. Methods of describing what the spectrum tells us about a sample's composition are explained. On a more practical note, the components of UV-Vis instruments are discussed in detail and useful information about sample preparation, sampling device selection, and routine maintenance are provided. Finally, a glossary of key terms and concepts and a selective bibliography are provided to guide the reader to other avenues of further discovery on the subject of UV-Vis spectroscopy.[1–11]

1.1.1 *Basic Theory*

1.1.1.1 *The Electromagnetic Spectrum*

The electromagnetic spectrum consists of many different types of radiation (energy): gamma, x-ray, ultraviolet (UV), visible (Vis), infrared (IR), microwave, and radio waves (Figures 1.1 and 1.2). Each type of radiation occupies its own region of the electromagnetic spectrum, and the major difference between the individual spectral regions is merely the energy of the photons involved. This energy is expressed in units of photon energy (eV), Frequency (Hz), wavenumber (cm^{-1}), or wavelength (nm).

With the possible exception of gamma, most people are familiar with these types of radiation through their experiences in daily living. X-rays penetrate our

Molecular Electromagnetic Spectrum Regions

Region	Wavelength (nanometers)	Wavenumbers (cm^{-1})	Frequency (Hz)	Energy (eV)
Visible	360-780	27,778-12,821	8.3e^{14}-3.8e^{14}	3.4-1.6
Ultraviolet	190-360	52,632-27,778	1.6e^{15}-8.3e^{14}	6.5-3.4
Vacuum UV	10-190	1e^{6}-5.26e^{4}	3.0e^{16}-1.6e^{15}	124-6.5

Fig. 1.1 The electromagnetic spectrum. The Ultraviolet range is from 190 through 360 nm.

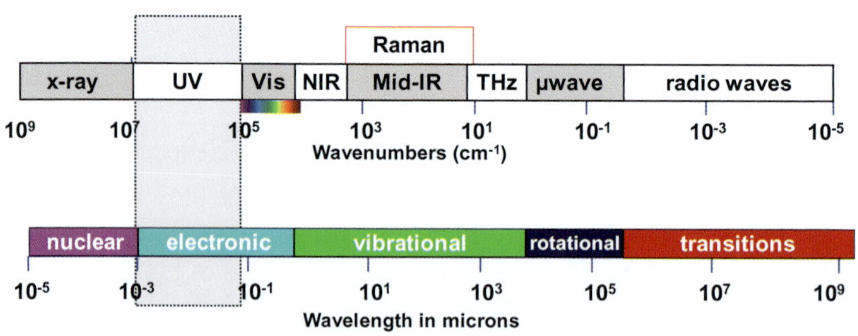

Fig. 1.2 Comparative regions of the electromagnetic spectrum (Ultraviolet region).

bodies, allowing physicians to visualize our internal anatomy. UV light is associated with sunburn and tanning. We see colors and objects in the visible spectrum. In the kitchen, we toast our bread with IR radiation, and "zap" or heat our meals with microwaves. We use microwave transmission for cell phone technology and we use radio waves to broadcast sound and images through AM/FM radio and television signals. The spectral region most useful in day-to-day analytical chemistry is the range of wavelengths from just below our visual perception (i.e., Ultraviolet or UV) through the colors we see (i.e., visible or Vis), known collectively as the UV/Vis spectral region. The wavelengths covered in the UV/Vis are measured in nanometers (nm), a unit of length representing one billionth of a meter (i.e., 10^{-9} m). The generally accepted ranges for the UV/Vis region are ultraviolet (190–380 nm) and visible (380–750 nm). Some UV/Vis work involves longer wavelengths in the neighboring shortwave near-IR spectral region, from about 750–950 nm.

1.1.1.2 *Molecular and Electronic Spectroscopy Unit Conversion*

The units used to describe spectra for molecular spectroscopy vary with the academic or engineering discipline describing the spectral regions. For the most part the spectra are described in terms of an optical response specific to energy from a particular spectral region. The responses are in terms of inelastic or elastic collisions of the energy with various materials and are referred to as scattering absorption, absorbance, reflection, transmission, and so on. The various spectral energy for molecular and electronic spectroscopy is described in terms of units of wavelength e.g., microns (μm) as 10^{-6} meters, or nanometers (nm) as 10^{-9} m, or Angstroms (Å) as 10^{-10} m. Energy is also expressed as frequency (cycles per second or sec.$^{-1}$) termed Hertz (Hz), as Wavenumbers (cm^{-1}), and as electron Volts (eV).

The various formulas used for the inter-conversion of units for spectroscopy include the following. Light has both particle and wave properties, quantum theory tells us that the energy of a light "particle" or photon Ep is given by:

$$E_p = h\nu \tag{1.1}$$

$$E_p = hc\bar{\nu} \tag{1.2}$$

where h = Planck's constant (or 6.6256×10^{-27} erg-sec.), and ν (nu) is the frequency of light, also known as Hertz (Hz) or the number of vibrations per second, in units of sec.$^{-1}$.

$$\bar{\nu} = \frac{\nu}{c \cdot n^{-1}} = \frac{1}{\lambda} \tag{1.3}$$

$$\nu = \frac{c \cdot n^{-1}}{\lambda} \tag{1.4}$$

Where $\bar{\nu}$ is the wavenumber units as cm^{-1} (i.e., the number of waves per centimeter); c is the velocity of light in a vacuum (or 2.9979×10^{10} cm·sec^{-1}); ν (Greek: nu) is the frequency of light; also known as Hertz (Hz) or the number of vibrations per second; in units of sec.$^{-1}$, n is the refractive index of the medium the light is passing through (e.g., air = 1.0003); and λ (Greek: lambda) is the designation for wavelength in units of nm.

Combining the above equations we also note:

$$\bar{\nu} = \frac{hc}{E_p} \Rightarrow \frac{1}{\lambda} = \frac{hc}{E_p} \Rightarrow \lambda = \frac{E_p}{hc} \quad (1.5)$$

1.1.1.3 *Unit Conversions*

Wavelength (in nanometers) to wavenumbers (cm^{-1}):

$$\bar{\nu}_{cm^{-1}} = \frac{10^7}{\lambda_{nm}} \quad (1.6)$$

Wavenumbers (cm^{-1}) to wavelength (in nanometers):

$$\lambda_{nm} = \frac{10^7}{\bar{\nu}_{cm^{-1}}} \quad (1.7)$$

Wavelength (in nanometers) to frequency (in Hz):

$$\nu_{Hz} = \frac{2.9979 \times 10^{17} nm \cdot \text{sec}^{-1}}{\lambda_{nm}} \quad (1.8)$$

Wavelength (in nanometers) convert to energy (in eV):

$$E_{p_{eV}} = \frac{1240}{\lambda_{nm}} \quad (1.9)$$

1.1.2 *Basic Spectroscopic Measurement Instruments (Optical Configurations)*

One of the basic issues of spectroscopic measurement is that physical matter in any state (i.e., liquids, gaseous, or solid substances) will respond or interact when exposed to any of these types of electromagnetic radiation. The atoms and molecules that make up the matter will become "excited" and will (among other phenomena) absorb and re-emit certain amounts of the radiation as it passes through or collides with the matter. In the earliest spectroscopic experiments, scientists discovered that each atom or molecule will react in precisely the same way every time it is exposed to a particular type of radiation, predictably absorbing, emitting, or remitting precisely the same amount of energy. They also realized that this behavior is unique — no two different types of atoms or molecules respond in exactly the same way. From here, these pioneers learned to measure and record the behavior of matter when exposed to electromagnetic radiation. The resulting plot of molecular or atomic behavior at various wavelengths is known as a *spectrum* (plural form: spectra). Serving as an individual "fingerprint", a substance's spectrum gives scientists a tool for positing its correct identity without knowing a priori what it is.

In the UV/Vis range, light is absorbed by matter in a very selective manner. If the energy associated with a particular wavelength of light corresponds to the energy needed to boost an electron from its "resting" position to a higher energy level, an electron absorbs the photon of light and jumps to a higher energy level. It immediately returns to its resting position, but the energy used to move the electron is unavailable to be measured by a spectrometer, causing the detector to register a lower energy. Because light absorbance in the UV/Vis range is predictable and reproducible, these wavelengths are invaluable tools in the analytical lab for chemical analysis or measurement of color. We use them to reveal the quantity of a particular substance in a mixture, what its identity might be, and when a chemical reaction in which it is involved is complete. Figures 1.3 through 1.7 show the optical configurations for various UV/Vis spectrometer systems.

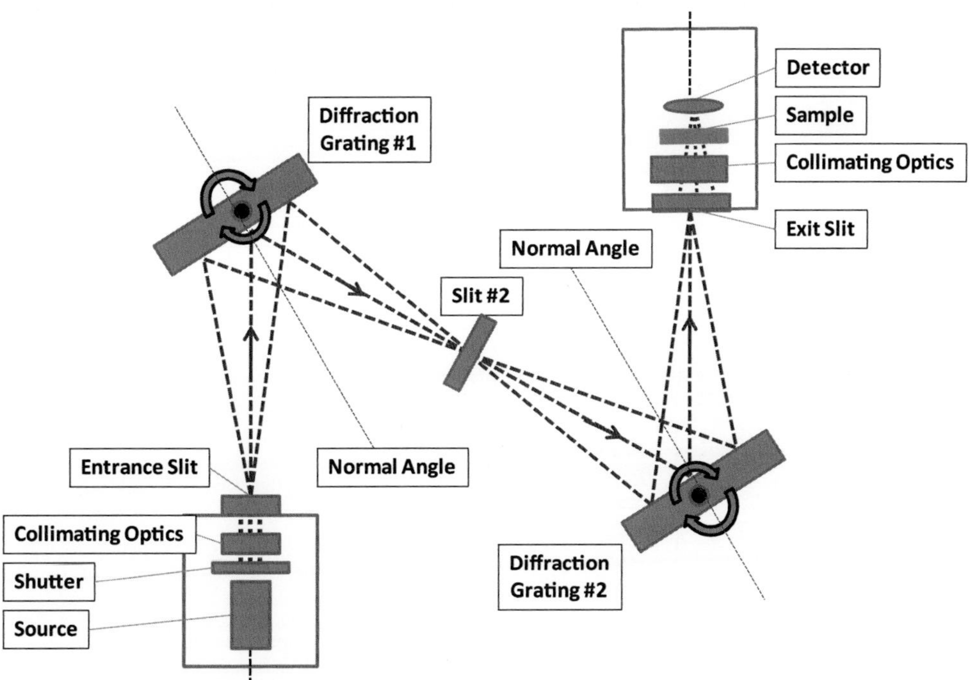

Fig. 1.3 Double monochromator optical system (dispersive-type).

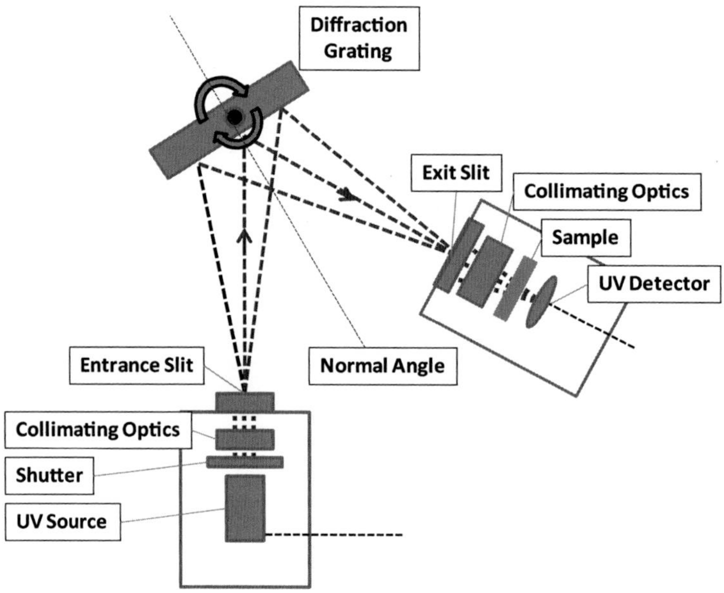

Fig. 1.4 Single monochromator optical system (dispersive-type).

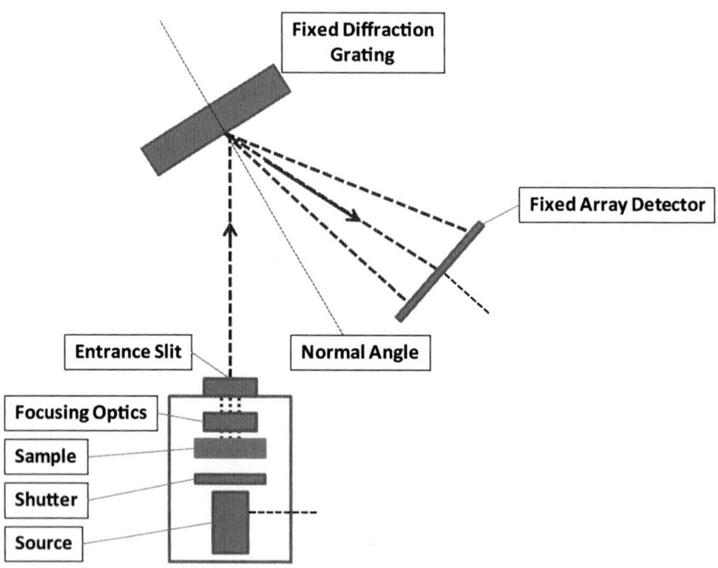

Fig. 1.5 Diode-array spectrograph optical system (dispersive-type).

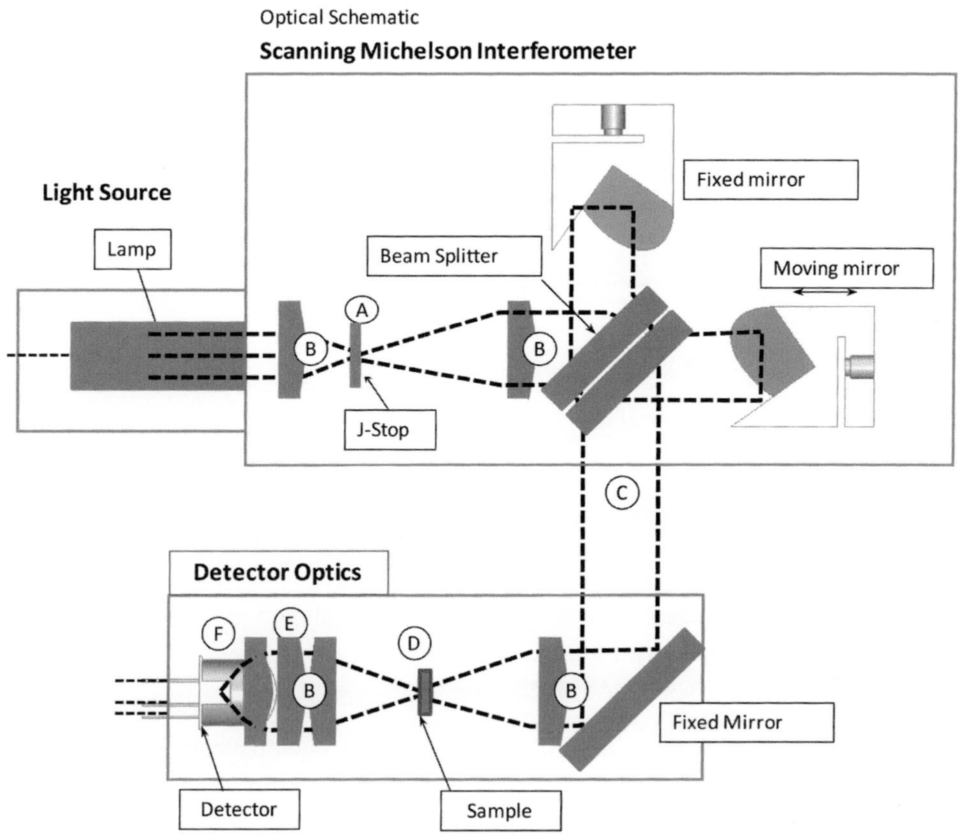

Fig. 1.6 Interferometer based optical system (Interferometer/FT-type). A = J-stop (aperture), B = collimating optics, C = optical train, D = sample cell, E = focusing optics, F = detector or Photomultiplier Tube (PMT).

1.1.3 *Components of a UV/Vis Spectrometer*

UV/Vis spectrometers share the same basic components, which can be configured to achieve specific optical effects (Figures 1.3 through 1.7). These components include a radiation source, a monochromator (wavelength selector), a sample holder, and a detector. The electronics include a basic power supply, computer board for processing data and for low-level control, a controller board for motor control functions, and a detector board for signal amplification and for analog signal to digital signal (i.e., A-to-D) conversion of basic instrument detector signals.

(a) Sources: The light used in UV spectrometry is usually produced by a deuterium lamp. Visible light is produced by incandescent or tungsten filament lamps similar

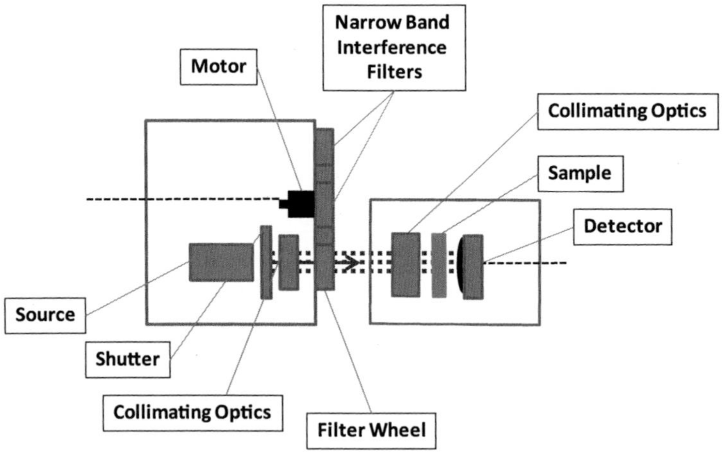

Fig. 1.7 Interference filter based photometer optical system (filter type).

to those used in household or projector bulbs. Common lamps and their working emissivity spectral ranges include quartz tungsten-halogen single filament lamp (220–2700 nm); DC deuterium lamp (185–375 nm); pulsed xenon arc lamp (180–2500 nm); and DC arc source (200–2500 nm).

(b) Wavelength Selectors: In order for the spectrometer/spectrophotometer to analyze the sample, the output of the light source must be broken into its individual wavelengths. Earlier instruments used quartz prisms to disperse light. Today, this is usually accomplished using a holographic diffraction grating, a device with a series of minute physical grooves on its surface to diffract the light as it strikes the surface of the grating. The polychromatic (many-colored) white light from the source is focused onto the grating. The precise grooves on the surface disperse the light according to its component wavelengths and the resulting spectrum (in the visible range) appears in the familiar "rainbow" color sequence; the ultraviolet energy is dispersed just before the visible violet light. The grating may be rotated along its axis to "aim" the desired wavelength onto the exit slit and then onto the sample for analysis. Multiple order light is dispersed from the grating so if more than one octave is measured a band filter must be incorporated to eliminate any orders of light other than the desired wavelengths. The diffraction grating will disperse light into the zero-order (mirror reflectance), first order dispersion, second-order dispersion, and even higher order dispersion (Figure 1.8).

(c) Sample Holders: The most common sample holders in UV/Vis analysis are cuvets (also spelled as cuvettes), which are fabricated from materials that do not interfere with the light passing through the cuvette body nor react chemically with the

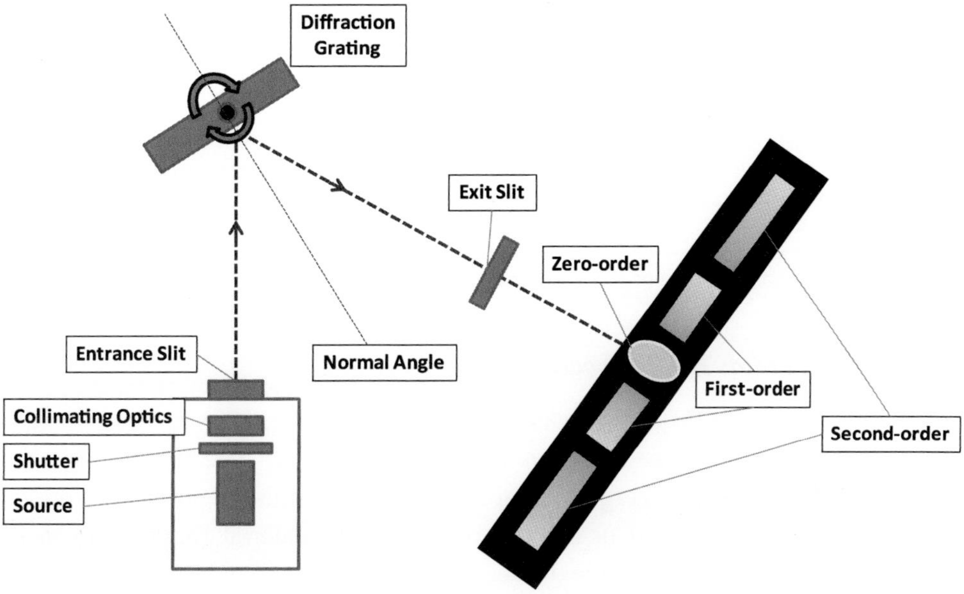

Fig. 1.8 Illustration of dispersion orders reflected/diffracted from a diffraction grating.

sample. These include certain polymers, such as poly(methyl methacrylate) or quartz with optically flat wall surfaces. Strictly speaking, for ultraviolet measurements, a UV-quartz material should be used for cuvette windows. This has low-OH content and has low entrapped air bubbles making it transparent to UV and with low scattering properties.

(d) Detectors: UV light transmitted through a sample is almost always detected using a photomultiplier tube (PMT). The same detector may be used for visible light, but far less expensive devices such as silicon diodes or photocells are also used. In these diode detectors, a chip of pure silicon is "doped" with the proper element such that an incoming photon causes a momentary electric impulse. The signals generated by these detectors are amplified and used to quantify the amount of light emerging from a cuvette. A more complete list of common detectors, common abbreviations, and their useful working wavelength ranges would include *silicon photodiodes* (350–1100 nm); *photomultiplier tubes* (PMTs, 160–1100 nm); *charge-coupled devices* (CCDs, 180–1100 nm); and *silicon-based photodiode arrays* (180–1100 nm).

(e) Basic Measurement Modes: The measurement modes for UV/Vis spectrophotometry are given in Table 1.1. A well designed spectrophotometer is capable of any of these measurements with slightly different configuration modes.

Table 1.1 Basic UV/Vis-near-IR measurement modes.

Instrument measurement mode	Description of measurement
Scan	Absorbance* vs. wavelength
Timedrive	Absorbance* vs. time at each wavelength
Individual wavelength(s) stop	Individual Absorbance(s)* at selected wavelength(s)
Quantitative methods	Concentration of analyte vs. absorbance
Kinetics	Kinetic rates of reaction

*As –log (transmittance) or –log (reflectance)

1.1.4 *The Concepts Behind Spectroscopy*

(a) Beer's Law: Beer's law is a mathematical way of expressing how light is absorbed by matter in a spectroscopic experiment. The law states that the amount of monochromatic light emerging from a sample is diminished by three physical phenomena: 1) the quantity or amount of absorbing material (molecules) in its pathlength (concentration); 2) the distance the light must travel through the sample (pathlength); and 3) the probability that the photon of that particular wavelength will be absorbed by the material (absorptivity or extinction coefficient).

This relationship may be expressed as:

$$A = \varepsilon bc \tag{1.10}$$

where $A =$ absorbance; $\varepsilon =$ molar extinction coefficient (i.e., absorptivity); $b =$ pathlength (in cm); and $c =$ molar concentration. This is referred to as the Beer-Lambert law or more simply Beer's law. There are many sections in this volume describing this concept in detail. Please refer to the index.

In practice, the absorbance is measured by comparing (or ratioing) the light emerging from a reference cell with the light emerging from a sample cell. The intensity of the incident radiation, I_0, is ratioed to the light emerging from a sample, I. The ratio I/I_0 is referred to as *transmittance*, or simply T. This can be directly plotted against the concentration, but the relationship is not linear. The negative log 10 of the transmittance, however, is linear with concentration. Therefore, absorbance is measured as: $-\log_{10}(I/I_0) = A = abc = \varepsilon bc$.

(b) Bandpass and Resolution: The terms *bandpass* and *resolution* are used to express the capability of a spectrometer/spectrophotometer to separate spectral bands or lines that are separated by some finite distance. For an instrument that disperses energy over a pre-specified region of the electromagnetic spectrum, the bandpass of a spectrometer is used to describe which portion of the spectrum can actually be isolated by the spectrometer in a "pure" wavelength form. The spectrometer bandpass is dependent upon the dispersion of the grating and the entrance and exit slit widths. An illustration is often used to show the problem associated with measuring monochromatic light using conventional spectrometers. If the ideal spectrometer were used to measure a bright line emission spectrum at a single wavelength (λ), the spectrum would appear as a single line (Figure 1.9). What really occurs when such a spectrum is measured using a conventional spectrometer is a broadband spectrum due to the width of the instrument function, as shown in Figure 1.10.

The measured line spectrum assumes a Gaussian-like (or bell-shaped) curve. This characteristic broadening of a line spectrum through the spectrometer is an illustration of the spectrum bandpass (aka bandwidth). The actual bandpass for any instrument is assigned a value by determining the full width at half

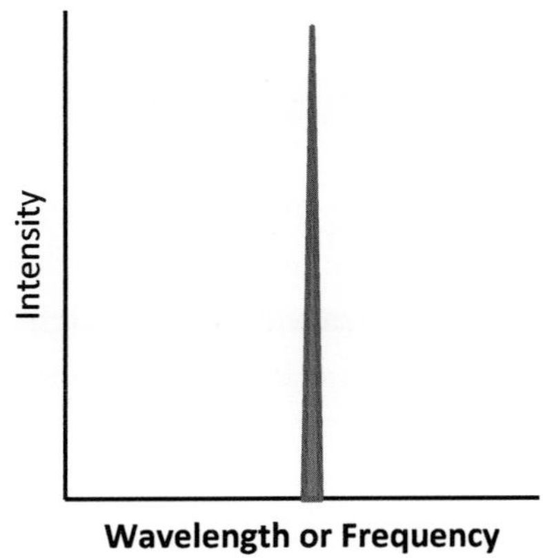

Fig. 1.9 The bright line emission spectrum at a single wavelength as it would appear in an ideal spectrophotometer.

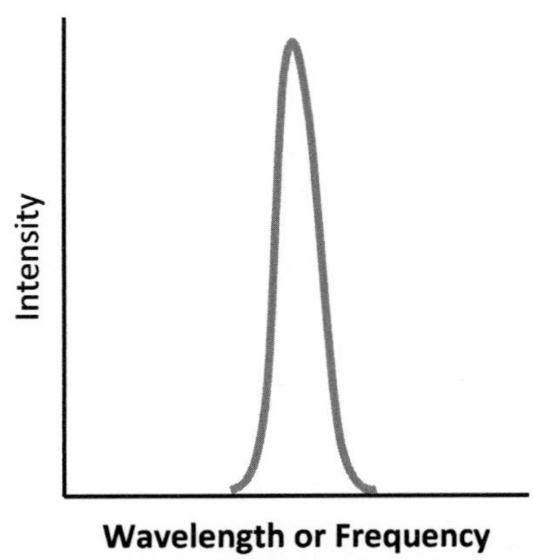

Fig. 1.10 Spectrum of a bright line emission source (e.g. deuterium lamp) from a standard spectrophotometer. The characteristic broadening is an illustration of the real bandpass/bandwidth of a spectrophotometer.

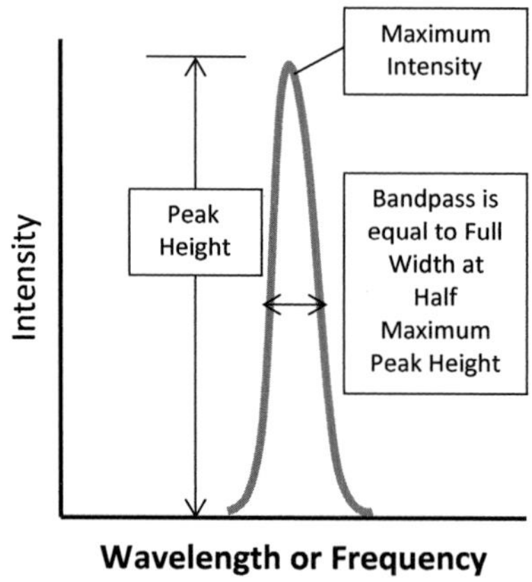

Fig. 1.11 Illustration of the determination of bandpass/bandwidth using the bell-shaped peak obtained from measuring the spectrum of a bright line source projected through a monochromator optical system.

maximum (FWHM) height of the bell-shaped spectrum. Thus, the FWHM could be empirically determined by finding the wavelength where maximum intensity occurs, and measuring the peak height at this position. This height measurement is divided in half and the bandwidth measured at this height on the band as illustrated in Figure 1.11. The actual shape of a band is the result of several instrumental characteristics including the overall quality of the optics and detector systems, as well as the width and positions of the entrance and exit slits relative to the dispersive element (e.g., diffraction grating).

The effective *resolution* of a spectrometer is the minimum distance between two adjacent peaks that can be detected by the spectrometer under designated operational performance settings. Resolution is manually estimated or calculated by multiplying the slit width (generally expressed in mm) by the reciprocal linear dispersion of the monochromator (in nm per mm). The actual resolution of a spectrometer must be slightly greater (wider or poorer) than the theoretical value because of practical issues such as non-ideal optics.

(c) Signal-to-Noise: Simply stated, the *signal-to-noise ratio* (S/N) in a spectrometer is the numerical ratio of the amount of electrical current generated by the light from the sample striking the detector (at 100% T) relative (ratioed) to the amount of background signal shown by the detector when no light from the sample is present (i.e., baseline noise). S/N is an important factor in determining the precision and accuracy of a UV/Vis instrument, especially when dealing with low levels (i.e., concentrations) of analyte or for highly absorbing samples.

(d) Dynamic Range: The range of a specified analyte concentration over which a sensor or spectrophotometer response is directly proportional to a change in

Table 1.2 Measurement error and maximum dynamic range associated with instrument stray light specification.

Stray light specification of instrument (in % T)	Absolute maximum dynamic range (in Au)	Percent measurement error for 1 Au sample
0.0001	6	0.0004
0.001	5	0.004
0.01	4	0.04
0.1	3	0.39
1	2	3.71
10	1	25.96

concentration is referred to the *dynamic range* of a spectrometer. Dynamic range is stray light and noise limited. When calculating the maximum theoretical dynamic range for an instrument, the relationship of stray light, and the maximum observable absorbance value requires knowledge of Beer's law and a few simple mathematical relationships. This relationship is described in detail within other chapters; please see index for stray light, or dynamic range.

(e) Stray Radiant Energy: Any measured radiation reaching the detector that has not interacted with the sample at the chosen wavelength qualifies as stray light. This problem can originate from physical "light leaks" allowing outside light to enter the sample compartment, from the scattering of light from optical components within the instrument, or scattered light through (or from) the sample itself. The accompanying Table 1.2 shows the relationship between an instrument stray light specification, the maximum dynamic range of the instrument, and the percent error of a measurement. Stray light measurements are made using a sharp cutoff optical filter as an optical material or liquid. Examples of such filters include saturated solutions of potassium ferromanganate or lithium carbonate. Other solutions with an abrupt cutoff feature include saturated aqueous solutions of potassium bromide, potassium chloride, potassium iodide, sodium nitrate, and pure acetone. Tables 1.5 (Chapter 1) and 7.3 (Chapter 7) show typical cutoff wavelength ranges for solvents used during ultraviolet measurements.

1.1.5 *The Use and Maintenance of UV/Vis Spectrometers*

1.1.5.1 *Applications of UV/Vis Spectroscopy*

The following section provides an overview of the application of UV/Vis spectroscopy for pure and applied research.

(a) Qualitative and Semi-quantitative Applications: The ability of UV energy to be absorbed in proportion to its concentration has found use in many industries, just a few of which are described here.

Pharmaceutical companies use UV/Vis spectroscopy in numerous applications: measuring the strength of any component in a dosage form; monitoring dissolution (dissolving from a tablet or capsule into solution); analyzing protein and DNA (See Chapter 7); identifying polymers; and detecting species separated by thin-layer (TLC) and high performance liquid chromatographic (HPLC) techniques.

Chemical companies use the UV and visible wavelengths to follow the progress of chemical and fermentation reactions, determine isomeric purity of products, and in titrations with a specific measureable absorbance reaction (endpoint determination).

Universities, technical schools, and other teaching environments use UV/Vis because it is the simplest and least expensive instrumental method with which to demonstrate spectroscopic analysis techniques.

(b) Quantitative Applications: For UV absorbing chromophores (see Chapters 7 and 8) the method is extremely useful for measuring quantities of specific molecules in solution. The basic UV quantitative method uses Beer's law for calculations of absorbance versus concentration at specific measurement wavelengths.

The types of electrons that can be excited by UV/Vis light are few in number: nonbonding electrons, electrons in single bonds, and electrons involved in double bonds. These may be excited to several excited states. The distinction between molecules is that the ability to "jump" to higher states is affected by attached moieties — for example, double bonds; conjugations; and elements such as oxygen, bromine, and others with pairs of nonbonding electrons. As a consequence, most elements with UV/Vis absorbances have specific wavelengths at which their peak absorbances occur. These peaks may be used to identify a particular molecule for quantitative analysis.

The UV/Vis region is not as "rich" in information as the infrared spectrum, but it reveals enough detail to enable a comparison of a material with a previously identified substance. A common use of this capability is in the pharmaceutical industry, where UV/Vis detectors are frequently used with HPLC instruments as a final check before a drug product is released for sale to consumers.

(c) Successful Applications of Ultraviolet Spectroscopy

Selection of Measurement Technique and Pathlength: Samples are analyzed by aiming light through a sample holder (for transmittance measurements) and by separating light into individual wavelengths as it passes through the sample. The term *pathlength* refers to the distance the light travels through the sample, as measured by the interior dimensions of the cuvette. The following pathlengths are recommended for analysis in UV/Vis through near-IR spectral ranges: **UV/Vis** (190–750 nm) 0.1 to 10 cm; **short-wave near-IR** (800–1100 nm) 5 to 10 cm; **long-wave near-IR** (1050–3000 nm) 0.1 to 2 cm.

Spectroscopists have the choice of measuring the light just as it is transmitted through the sample, *transmission* or *transmittance* (straight and diffuse); as it is reflected off the sample, *reflection* or *reflectance* (specular and diffuse); or a combination of these effects, such as *transflectance* or *interactance.*

The physical nature and appearance of the sample should determine which of these geometries is employed. Samples can possess any of several different physical or optical properties that affect analysis. The following are basic guidelines for selecting a sample presentation geometry based upon the optical properties of each sample.

- *Clear solids (optical materials)*: Transmittance
- *Translucent or opaque solids*: Diffuse reflectance or diffuse transmittance (for turbid samples)
- *Reflecting (mirror-like) optical surfaces*: Specular reflectance
- *Clear liquids:* Transmittance
- *Translucent or opaque liquids*: Reflectance or diffuse transmittance
- *High optical density (highly absorbing samples)*: Tiny pathlengths in transmittance

The accompanying Tables 1.3 and 1.4 list the characteristics of common cuvettes, including the materials they are made of, their dimensions, and other criteria for selecting the proper cell for the sample and measurement technique at hand.

1.1.5.2 *Calibration and Basic Maintenance*

General Concepts of Calibration: Because it is difficult to predict the exact response of an instrument to a material at a given concentration, virtually all calibrations are empirical. An empirical calibration involves the preparation of multiple samples in several concentrations across the expected concentration range for test samples. A calibration curve (or straight line) is built by graphing the response of the instrument (generally as absorbance) to the concentration of the material (i.e., weight per unit volume). A wavelength is chosen at which the material has a large, distinctive maximum absorbance peak. Such a graph (or spectrum) can be used to quantify unknown concentrations of later test samples.

Table 1.3 Characteristics of UV-Vis window/cuvette materials.

Optical material	Transmittance range (nm)	Refractive index (at 600 nm)	Relative rupture strength (sapphire = 100)
Poly(methylmethacrylate)	250–1100	1.4906	4.3
UV-grade fused silica	200–2500	1.4580	10.9
Synthetic fused silica	230–2500	1.4580	10.9
Crystalline quartz (SiO_2)	240–2500	1.5437	2.3
Quartz, extremely low O-H	190–2500	1.5437	2.3
Flint glass (SF 10)	380–2350	1.7268	3.8
Flint glass (SF 8)	355–2350	1.6878	3.7
BK 7 glass	315–2350	1.5165	3.7
Optical crown glass	320–2300	1.5226	3.7
Borosilicate crown glass	360–2350	1.4736	3.7
Pyrex	360–2350	1.4736	3.8
Tempax	360–2350	1.4736	3.8
Sapphire (Al_2O_3)	150–5000	1.7677	100
Sodium chloride	250 nm-16 m	1.5400	0.5
Suprasil 300	190–3600	1.5400	3.8
Diamond	220–4000	2.4000	83.7
Spectrosil	170–2700	1.5400	3.8
Infrasil	220–3800	1.5400	3.8

Table 1.4 Typical sampling accessories and dimensions for UV-Vis Spectroscopy.

Cuvette type	Outer dimensions (mm)	Pathlength (mm)	Capacity (mL)
Transmittance (1 mm pathlength)	45(H) × 12.5(W) × 3.5(L)	1.0	0.3
Transmittance (5 mm pathlength)	45(H) × 12.5(W) × 7.5(L)	5.0 (0.5 cm)	1.5
Transmittance (10 mm pathlength)	45(H) × 12.5(W) × 12.5(L)	10.0 (1 cm)	3.0
Semi-micro cell	45(H) × 12.5(W) × 12.5(L)	10.0 (1 cm)	1.0 or 1.5
Micro cell	25(H) × 12.5(W) × 12.5(L)	10.0 (1 cm)	0.5
Cylindrical cell (10 mm pathlength)	10(L) × 22(D)	10.0 (1 cm)	3.1
Cylindrical cell (10 mm pathlength)	20(L) × 22(D)	20.0 (2 cm)	6.3
Cylindrical cell (10 mm pathlength)	50(L) × 22(D)	50.0 (5 cm)	16
Cylindrical cell (10 mm pathlength)	100(L) × 22(D)	100.0 (10 cm)	31
Micro flow cell	50(L) × 12.5(W) × 12.5(L)	10 (1 cm)	0.4 or 0.6
Test tube (round)	75(H) × 12(D)	10 (1 cm)	5.9
Test tube (round)	105(H) × 19(D)	17 (1.7 cm)	23.8
Test tube (round)	150(H) × 19(D)	17 (1.7 cm)	34.0

Note: The Z dimension for a cuvette is the distance from the bottom of the instrument cell compartment to the center of the light beam, by convention this is typically 8.5 mm or 15 mm.

Photometric Accuracy: The National Institute of Standards and Technology (NIST) provides materials designed for reference measurements to verify photometric accuracy of spectrophotometers. The materials are referred to as Standard Reference Materials (SRMs). One well-known set of standards for UV/Vis analysis is SRM 930D, consisting of a set of three individual neutral density glass filters provided in standard cuvette size, black metal holders at nominal transmittances of 10, 20, and 30%. Photometric accuracy is a term used to describe how accurately a given spectrophotometer is able to measure the total transmittance of each individual sample in a set of SRMs. The method of making measurements of SRMs (or other well-characterized reference standards) has been delineated by the American Society for Testing and Materials International (ASTM International) in various documents. Many of these documents are referenced in this set of volumes.

Examples of other materials used for photometric accuracy measurements include SRM 931 (solutions of nickel and cobalt in nitric and perchloric acids) and SRM 2031 (metal on quartz with nominal transmittances of 10, 30, and 90%). For UV, SRM 935 (potassium dichromate in perchloric acid) and SRM 84 (potassium acid phthalate in perchloric acid) have been used. The reader is referred to Chapter 8 in this volume for details of reference materials used for ultraviolet spectroscopy.

Linearity Checks: The linearity of any instrument may be checked by using neutral-density interference filters. These are nonspecific (i.e., they cover multiple wavelengths) optical filters that reduce the amount of light by specified amounts, for example, by 10% T, 20% T, and so forth. If there is a problem with the linearity, the instrument manufacturer should be contacted. These checks may be made anytime, but should be performed at least once a year or before critical experiments. A calendar of instrument qualification tests should be maintained for quality assurance purposes.

Wavelength Accuracy Checks: If an instrument is not indicating the correct wavelength during an analysis, a significant measurement error will be generated. Solutions of known composition and purity are used to check wavelength accuracy. Examples of materials used for wavelength accuracy checks include: potassium dichromate at pH 2.9 (maximum absorbances at 257 and 350 nm, minima at 235 and 313 nm); holmium oxide glass filter (used for calibrations at 279.3, 333.8, 385.8, 536.4, or 637.5 nm); didymium oxide glass filter (used from 250 to 2000 nm); and samarium percholorate solution (used from 225 to 520 nm). Other wavelength check materials are demonstrated in Chapter 8 of this volume.

Instrument Care: The proper procedure for handling a UV/Vis spectrometer is prescribed by the manufacturer. The purchasing institution is responsible for assigning a staff member to clean and maintain each instrument. An electronic and hard copy log book is advised for any instrument in order to track lamp life, test methods, and repair frequency.

Why and How to Clean Cuvettes: Keeping cuvettes clean and organized contributes greatly to accurate and efficient analysis. Dirty and mismatched cuvettes are a common source of error. For a light cleaning, use a detergent wash, followed by multiple pure-water rinses. For heavier cleaning, repeat the previous step followed by cleaning with chromic-sulfuric acid solution wash and multiple pure-water rinses. When the cuvette is emptied, it is good technique to touch the inverted cuvette to a clean paper towel before righting it for the next sample. The drop or two of moisture that may run down the face of a cuvette will cause concentration (added dilution) or light scattering problems (staining).

Sample Preparation and Handling:

Solvents and Operating Conditions: Choosing the proper solvent is important for several reasons: 1) because the solvent must interact with the analyte to dissolve it, it has an effect on the spectrum, and 2) the solvent itself might have an absorbance at the wavelength used for analysis. Good solvents have their spectral properties delineated on the label: "Spectro-Grade" or "UV-Grade", and so forth. HPLC solvents may not be sufficient for sensitive UV spectroscopic work. Select a solvent that dissolves the sample but does not react with it. Water is notorious for its many chemical reactions. Acids and bases added (by the producer) to preserve a solvent might react with the analyte. Read the solvent specification sheet carefully taking care that additives are not included in the liquid which may interfere with UV measurements. The accompanying Table 1.5 lists typical UV solvents and their appropriate cutoff wavelengths.

Table 1.5 Typical UV solvents and appropriate UV cutoff wavelengths.

Solvent	UV cutoff (nm)
Acetonitrile	190
Water	190
Cyclohexane	195
Isooctane	195
n-Hexane	201
Ethanol (95 vol. %)	205
Methanol	205
Trimethyl phosphate	210
Acetone	220
Chloroform	240
Xylene	280

Table 1.6 Absorptions of UV chromophores (160–210 nm).

Chromophore	Absorption band location (nm)
Nitriles (R-C≡N)	160*
Acetylenes (-C≡ C-)	170*
Alkenes (>C=C<)	175*
Alcohols (R-OH)	180 (175–200)*
Ethers (R-O-R)	180*
Ketones (R-C=O -R′)	180*, 280
Amines, primary (R-NH_2)	190* (200–220)
Aldehydes (R-C=O –H)	190*, 290
Carboxylic acids (R-C=O –OH)	205
Esters (R-C=O –OR′)	205
Amides, primary (R-C=O –NH_2)	210
Thiols (R-SH)	210
Nitrites (R-NO_2)	271
Azo-group (R-N=N-R′)	340

**Note*: Absorptions below the cutoff for common solvents would not be observed in solvent solution measurements.

Chromophores: The grouping of atoms producing a characteristic absorption is called a *chromophore,* (*chromo* = color and *phore* = carrier). A specific grouping of atoms produces a characteristic absorption band at a specific wavelength. The intensity and location of these absorption bands will change with structural changes in the group of atoms and with solvent changes. The location of bands associated with UV chromophores is shown in Table 1.6.

Troubleshooting: Because of instrument complexities, especially electronics, mechanical precision, and optics; usually only limited maintenance can be performed directly by the operator or user. When a problem does occur, its cause may be isolated to determine if the problem is instrument or sample related. One may isolate these variables by measuring additional samples or by taking the same sample measurements on a different instrument.

Sample/Solvent Problem: If a sample or solvent problem is suspected, first try running the sample on a second instrument if available. It is always quicker to change instruments than to remake the standards and samples. If the problem persists on a different instrument, then the chemistry must be checked and evaluated. Remake one or two standards or samples using a different lot of solvent, using freshly cleaned glassware. If problems such as spectral changes or nonlinearity continue, consider a different type of solvent. Due to the basic premise that most commercial spectrometers are quite stable, the problem is often found in the sample preparation or handling or in unwanted chemical reactions with the solvent and sample material(s).

Instrument Problems: If it has been determined that the samples are correctly made, then the instrument should be checked. Other than any special steps recommended by the manufacturer, the linearity and wavelength checks mentioned earlier would be useful to complete. If the problem is not easily solved, the best thing to do is call the instrument company for help rather than try to service the instrument yourself.

Avoiding Problems: Good laboratory techniques and reproducible sample work-up are good habits to develop. Each step of an analysis must be clearly stated and adhered to. Note that routine maintenance is an important aspect for avoiding problems. Maintaining an electronic and hard copy logbook for each instrument allows the analyst to determine whether a source lamp needs to be replaced or whether the detector may be aging. Cleaning is also quite important. Fumes and spills in and around the instrument will eventually destroy or cloud mirrors, gratings, and lenses; damage electronics or foul mechanical parts.

1.1.6 *Mini-glossary*

Absorbance: Measure of concentration of material present: Negative log (base 10) of Transmittance $[-\log_{10} T]$ as product of extinction coefficient (absorptivity), pathlength, and concentration, written as $A = abc = \varepsilon bc$.

Absorptivity: Probability of light absorbing at a particular wavelength for a specific analyte under specific conditions, e.g., pH, solvent, and temperature. Thus, a specific amount of material at specified conditions will absorb a specific fraction of the light striking it. Absorptivity is signified by an epsilon (ε) or a small "a".

Analyte: The particular material or quality to be determined in an analytical analysis.

Beer's Law: Relationship between the amount of light absorbed by an analyte and its concentration (c), pathlength (b), and absorptivity (ε or a, expressed in mass per unit volume or as molarity), written as $A = abc$.

Calibration: Determining the response of some analytical method to known amounts of pure analyte.

Calibration Curve (Line): The results of a calibration when graphed, usually Cartesian coordinates, e.g., concentration (in molarity) versus absorbance at a specified wavelength (in microns or nanometers).

Concentration: The amount of analyte in a given volume of solvent solution, e.g., moles per liter.

Cuvette/Cuvet: Transparent receptacle in which sample solutions are introduced into the light path of spectrometers. Usually, two sides are equal, i.e., 1 cm × 1 cm, while the third dimension is elongated, possibly as long as 15 cm. For UV work, the material is low-OH, high quality, quartz. Visible work permits use of glass or plastic cuvettes.

Detector: Device used to detect the intensity of the radiation from either sample or reference beams. Usually a simple silicon diode or a more sensitive photomultiplier tube.

Extinction/Extinction Coefficient: Words sometimes substituted for absorptivity.

Frequency: The number of times per unit length that the magnitude of an electromagnetic wave traverses from maximum to minimum then back to maximum amplitude. The unit for the number of waves per second is Hertz (Hz).

Grating: A reflective surface covered with evenly spaced, microscopic grooves, whose purpose is to separate individual wavelengths from "white" light or a broad band source into discrete wavelengths. The distance between grooves and the angle of the faces are determined by the wavelengths to be separated. The grating (except for diode arrays) is rotated at a known angle relative to the source and detector and the desired wavelength is emitted through an exit slit onto the sample or standard and then detected.

Limit of Detection: Lowest amount of analyte that may be consistently detected above the noise level of instrument.

Linear: A straight line. In context, this means that for a doubling of the analyte concentration, the signal will double. This allows prediction of unknown concentration from a calibration curve (or line).

Linearity: An experiment demonstrating that the response of an instrument changes in a predictable and linear fashion with increasing analyte concentration.

Noise: Any signal generated by the detector not directly responding to the light transmitted at the required wavelength.

Pathlength: The distance the light passes through the sample in its holder. In practical terms, the inside dimension of the cuvette.

Reference: In context, everything that is in the sample light path except the analyte of interest. This includes the cuvette, solvent, and any buffer or matrix used to prepare the sample.

Reference Standard: Analyte wherein the purity is documented and used to construct a calibration curve/line.

Scanning: The process where the wavelength range of the system is viewed in order, usually from lowest to highest wavelength. This usually occurs when the grating is rotated about its axis in a precise and measured manner.

Sensitivity: See "Limits of Detection".

Signal: The output of the detector due to its response to light emerging from the sample within the sample holder or reference cell.

Signal-to-Noise Ratio: The numerical ratio of the total signal at 100% T to the standard deviation of the variation of the instrument with no signal, i.e., the noise of the instrument.

Slit Width: The size of opening for the slit through which light passes. Size depends on wavelength range, separation ability of the wavelength selector (e.g., grating), and desired isolation of specific wavelengths (i.e., bandpass or resolution). This dimension of an instrument is either fixed or automatically programmed.

Solvent: Liquid used to dissolve the sample analyzed. Commonly water or methanol, both of high purity. Usually designated as specially purified for UV work, e.g., "Spectro-Quality" or "Spectro-Grade". Water must have high resistivity due to low concentrations of ions (resistivity of water should be greater than $18.0\,\mathrm{m}\,\Omega \cdot \mathrm{cm}$ at 25°C).

Solvent Cutoff: The wavelength at which the solvent absorbs a significant portion of the light, causing a loss of signal and inability to perform an analysis. In other words, the solvent becomes opaque to the wavelengths being used. This is common in the ultraviolet, rare in the visible (See Tables 1.5 and 7.3 of this volume).

Source: Also known as the "lamp". This is the origin of the light used in the spectrometer, an incandescent wire for visible light, a deuterium gas discharge for ultraviolet.

Spectrum: Series of wavelengths of radiation, belonging to a specific portion of the electromagnetic continuum, e.g., the ultraviolet spectrum, where the wavelength scale is defined as approximately 190 nm to 360 nm (or to 400 nm in some cases).

Stray Light: Any radiation reaching the detector that is not remitted from the sample at the chosen wavelength.

Ultraviolet: The portion of the electromagnetic spectrum below the violet in wavelength or above the violet in frequency (or energy). It is not detectable by the human eye. The portion of the spectrum approximately from 190 nm to 360 nm.

Wavelength: The distance from one crest of an electromagnetic wave to the same position on the subsequent wave. Peak-to-peak distance, usually measured in nanometers or microns.

References

1. R. C. Denney and R. Sinclair, *Visible and Ultraviolet Spectroscopy* (John Wiley & Sons, New York, 1987).
2. H.-H. Perkampus, *UV/Vis Spectroscopy and Its Applications* (Springer-Verlag, New York, 1992).
3. J.W. Robinson, *Undergraduate Instrumental Analysis,* 4th ed. (Marcel Dekker, New York, 1989).
4. R.A. Sawyer, *Experimental Spectroscopy* (Dover Publications, New York, 1963).
5. J.R. Edisbury, *Practical Hints on Absorption Spectrometry: Ultraviolet and Visible* (Plenum Press, New York, 1967).
   G.A. Vanasse, Ed., *Spectrometric Techniques — Vol. II* (Academic Press, Boston, 1981).
6. A. Knowles and C. Burgess, Eds., *Practical Absorption Spectrometry Techniques in Visible and Ultraviolet Spectrometry, Vol. 3* (Chapman and Hall. London, 1984).
7. C. Burgess and D.G. Jones, Eds., *Spectrophotometry, Luminescence, and Colour: Science and Compliance* (Elsevier Science B.V., Amsterdam, 1995).
8. R.M. Silverstein, G. Clayton Bassler, and Terence C. Morrill, *Spectrometric Identification of Organic Compounds,* 5th ed. (John Wiley & Sons, New York, 1991).
   ASTM Practices for UV/Vis Spectroscopy:
9. ASTM E169-04 (2009), Standard Practices for General Techniques of Ultraviolet-Visible Quantitative Analysis, Annual *Book* of *ASTM Standards*, Volume 03.06, 2009. ASTM International, 100 Barr Harbor Drive, PO Box C700, West Conshohocken, PA, 19428-2959 USA.
10. ASTM E275-08 (2013), Standard Practice for Describing and Measuring Performance of Ultraviolet and Visible Spectrophotometers Annual *Book* of *ASTM Standards*, Volume 03.06, 2013. ASTM International, 100 Barr Harbor Drive, PO Box C700, West Conshohocken, PA, 19428-2959 USA.
11. ASTM 958-13 (2013), Standard Practice for Estimation of the Spectral Bandwidth of Ultraviolet-Visible Spectrophotometers, Annual *Book* of *ASTM Standards*, Volume 03.06, 2013. ASTM International, 100 Barr Harbor Drive, PO Box C700, West Conshohocken, PA, 19428-2959 USA.

Chapter 2

Theory of Ultraviolet Spectroscopy, Light Interaction with Matter

Contents

2.1 Theory of Spectroscopy: Light Interaction with Materials

Spectroscopy refers to the interaction of electromagnetic energy with matter, using the science of spectrophotometry [Greek: (*specere* = to look at) + (*photos* = light) + (*metron* = to measure)]. The science of measuring ultraviolet energy interaction with matter as absorption, reflection, and scattering is the subject of this chapter. The matter interacting with ultraviolet energy is comprised of liquids, gases, semi-solids, slurries, solid composites, films, and ground particles; and is defined as the qualitative or quantitative measurement of the interaction of electromagnetic radiation (emr), as ultraviolet energy, with the particles themselves and with the atoms or molecules within a defined volume of matter.

The emr interacts mainly as scattering, absorption, reflectance, and emission when "confronting" solid matter. A variety of spectrometer configurations are used to optimize the measurement of electromagnetic radiation as it interacts with matter. The subject of this chapter is a broad overview of the different ways light, as particles and waves, interacts with molecules. This chapter is intended to be a reference chapter for referring to the various characteristics and mathematical descriptions of basic light interactions with atoms and molecules. Note: some portions of this chapter are reproduced from J. Workman and A. Springsteen (Eds.), Applied Spectroscopy: A Compact Reference for Practitioners, Academic Press, Boston (1998). (Used with Permission).

2.1.1 *The Physics of Light Interaction with Solid Materials*

Light interacts with solid materials as scattering, absorption, transmission (transmittance), reflectance (both regular and diffuse reflection), and diffraction. The purpose of spectroscopy is to quantify or qualify these interactions by the use of a variety of photon producing and photon detection devices. Some of the physics of these interaction phenomena will be presented within this chapter. A general reference describing this theory in more detail is found in reference.[1]

Solids, as black body radiators, emit light and this emitted light can be characterized by its radiated power, spectral profile, and photon flux. The spectroscopy of solids is a vast field which would encompass many hundreds of volumes and thus the scope of this chapter is limited to the basic mathematical concepts and relationships related to the transmittance/transmission or reflectance/reflection interaction as molecular spectroscopy of liquids and solid materials.[2–8]

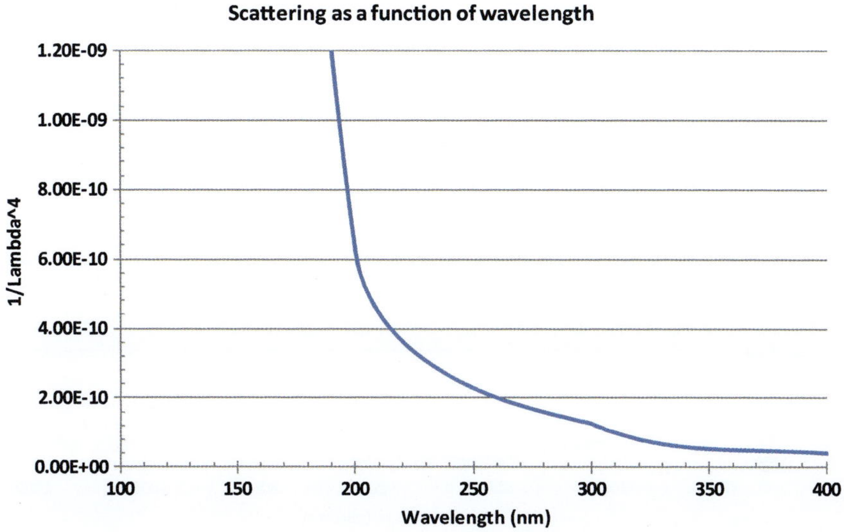

Fig. 2.1 Scattering as a function of wavelength to 400 (in nanometers).

2.1.1.1 *Light Scattering*

Light scattering as defined by Lord Rayleigh is elastic scattered light, and occurs when the intensity of the scattered energy (I_{RS}) is inversely proportional to the 4th power of the incident light wavelength (λ), as given by Equation 2.1 and shown in Figure 2.1.

$$I_{RS} \propto \frac{1}{\lambda^4} \tag{2.1}$$

The intensity ratio of Rayleigh scattered light incident onto a solid particle surface is given by Equation 2.2.

$$\frac{\bar{I}_S}{I_0} = \frac{1}{r^2}\left(\frac{2\pi}{\lambda}\right)^4 \alpha^2 \sin^2\theta \tag{2.2}$$

Where $\bar{I}_S$ = the scattered radiation intensity flux per unit time (seconds); I_0 = the incident beam radiation intensity (in energy counts); r = the average particle radius; λ = the wavelength of the incident radiation; α = the proportionality factor (if a surface is isotropic, the α is by definition identical for all angles of reflectance); and θ = the reflected light angle versus normal incidence.[9–13] Note from Equation 2.2 that the scattered radiation intensity is proportional to the inverse of the square of

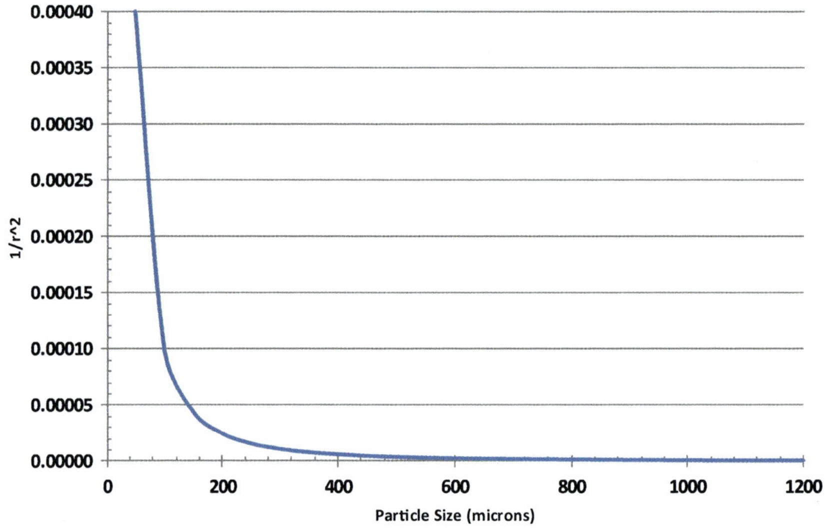

Fig. 2.2 Rayleigh scattering as a function of average particle size (radius) up to 1200 microns (1.2 millimeters).

the average particle size radius given as illustrated in Equation 2.3 and Figure 2.2.

$$I_s \propto \frac{1}{r^2} \tag{2.3}$$

2.1.1.2 *Absorption and Transmittance*

Light passing through an absorbing material is attenuated as a function of the absorption coefficient (a, or often denoted as ε) in units of Liter $\cdot$ Mole^{-1} $\cdot$ cm^{-1} of the material, and the thickness (t) of the material in cm following the relationship[14–15] as:

$$I = I_0 e^{-at} \tag{2.4}$$

Furthermore, the fraction (in T units) of a collimated beam of light of wavelength (λ) in nm transmitted through a material of index of refraction (n) and a thickness (t) in cm is given by:

$$\mathrm{I_T} = \frac{\mathrm{I_0}}{1 + \frac{4\rho^2}{(1-\rho^2)} \sin^2\left(\frac{2\pi \mathrm{nt}}{\lambda}\right)} \tag{2.5}$$

where $\rho^2 = \mathrm{R}$ (the reflectance/reflection) given for normal incidence as:

$$R = \frac{(n-1)^2}{(n+1)^2} \tag{2.6}$$

2.1.1.3 *Reflectance*

The reflectance (R) of light at normal incidence passing through a refractive index (n_1) into a second material of refractive index (n_2) is given by the equation:

$$R = \frac{(n_2 - n_1)^2}{(n_2 + n_1)^2} \tag{2.7}$$

For films or coatings of refractive index (n_f) deposited on glass with a refractive index of (n_g) as is the case with A-R (anti-reflective) coated optics, the reflection (R) from the coated surface is calculated using a simplified formula as:

$$R = \frac{(n_g - n_f^2)^2}{(n_g + n_f^2)^2} \tag{2.8}$$

The desired reflectivity may be computed using this equation.

2.1.1.4 *Optical Density*

The Absorbance (A or Au), or optical density (O.D.) of a material is defined by the mathematical relationship:

$$A = O.D. = -\log_{10}\left(\frac{I}{I_0}\right) = \frac{1}{2.303}\ln\left(\frac{I_0}{I}\right) \tag{2.9}$$

where I_0 = the intensity of the incident light beam, and I = the intensity of the transmitted beam after sample interaction. The O.D. of the material in which the light is passing through is given by the above expression and is a term used synonymously with absorbance.

2.1.1.5 *Opacity*

Is an expression that refers to the fraction of energy that is not transmitted by a sample, and is expressed using the equation:

$$\mathrm{O} = \frac{\mathrm{I_0}}{\mathrm{I}} \tag{2.10}$$

2.1.1.6 *Diffraction*

Diffraction can be loosely defined as the bending of light around objects.[16–18] The defining principle for diffraction is Christiaan Huygens' principle, which informs us that every point of a wave front can be considered a secondary point from which waves propagate in multiple directions. A single narrow slit can demonstrate the alternating light and dark patterns shown by diffraction.

The secondary waves propagated during diffraction have an intensity proportional to:

$$1 + \cos\frac{\theta}{2} \tag{2.11}$$

where θ is the propagation angle of the secondary wave in relationship to the original (or primary) wave. This relationship defines the intensity for waves propagated in the direction of the original wave as 1, and 0 intensity for the wave propagated at 90° to the original wave. The central bright band in a diffraction pattern is approximately 20 (or more) times brighter than the next (or adjacent) bright band in a typical diffraction pattern resulting from a narrow slit. The reader is referred to the index for additional information and treatment of the topic of diffraction and diffraction gratings.

2.1.1.7 *Emissivity*

The power (p) radiated or emitted by a black body radiator is given by the relationship:

$$p = kT^4 \tag{2.12}$$

where k is the Stefan-Boltzmann law proportionality constant equal to 5.67×10^{-12} Watts/($\text{cm}^2 \times {}^\circ\text{K}^4$), and T = °K = (°C + 273).

The spectral profile of this emitted radiation is given by Max Planck's hypothesis as shown in the following Equations 2.13 through 2.15.

$$E_p = h\nu \tag{2.13}$$

Where E_p = the photon energy (in joules) at a given frequency (ν, in units of sec.^{-1}), and where h = Planck's constant = 6.6256×10^{-3} joule-sec. (equivalent to 6.6256×10^{-27} erg-sec.). And since $\nu = \frac{c}{\lambda}$ it follows:

$$E_p = \frac{hc}{\lambda} \tag{2.14}$$

Thus, using this equation the energy of each photon at a particular wavelength is determined.

Furthermore, since 1 Watt = $\frac{\text{joule}}{\text{sec.}}$, the number of photons (N_p) per second in an emission phenomenon is calculated as:

$$N_p = \frac{1}{E_p} \tag{2.15}$$

where N_p = the number of photons per second (as the photon flux). Thus emission light sources are generated by heating metal filaments or metal surfaces to obtain the appropriate emission spectral characteristics in power, spectral profile, and photon flux.[4,7,19] These equations may be used to compose the requirements for an emission source based on the temperature of the source and the photon flux. Equations 2.13 and 2.14 may be used to calculate the spectral energy output relative to frequency or wavelength for a black body radiator. More detail is discussed for various types of emission sources, and the reader is referred to the index for Planck's constant and emissivity. (See also Volume 2, Chapter 2.)

2.1.2 *Spectrophotometry of Solids*

2.1.2.1 *Photometers*

Photometers are spectrophotometers (spectrometers) characterized as instruments which use interference filters to select the wavelength from a broad energy source that is made incident to the sample specimen. Photometers may be simple with anywhere from 1 to 3 wavelengths or more complex with multiple wavelengths and complex filter systems (up to 20 or more individual filters). Several such designs are illustrated in the section on instrumentation. The reader is referred to the index for Photometers. The optical design for a simple photometer is shown in Figure 2.3.

2.1.2.2 *Dispersive Spectrometers*

Dispersive spectrometers (or spectrophotometers) rely on the use of a monochromator (or monochromators) to disperse specific frequencies of light from a broadly emitting light source incident to the sample specimen. There is much

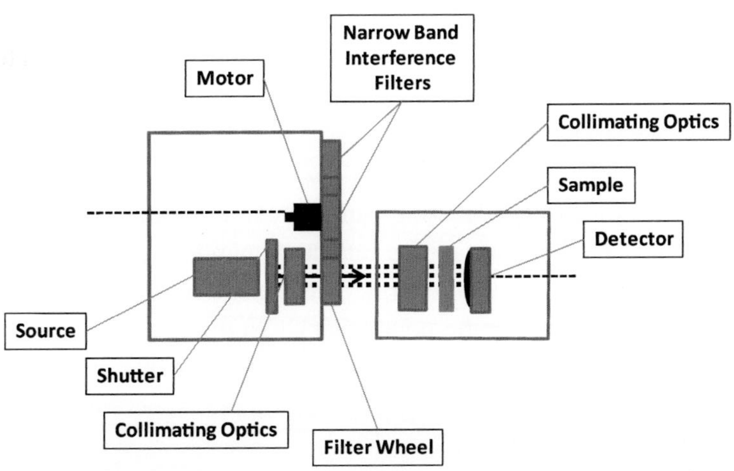

Fig. 2.3 Illustration of a simple filter photometer optical design.

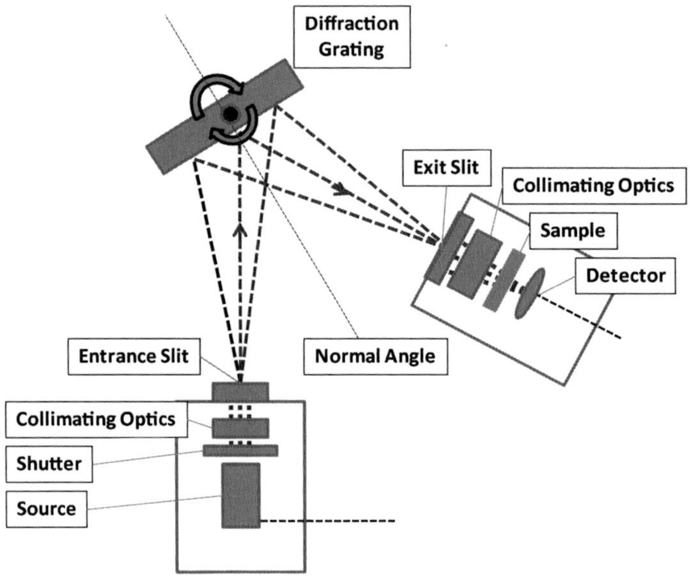

Fig. 2.4 Simple single monochromator optical design.

more information on this type of instrument found in the index under dispersive instruments or monochromators. A simple, single monochromator optical design is shown in Figure 2.4.

2.1.2.3 *The Integrating Sphere*

The purpose of an integrating sphere detector system is to provide a collection device for reflected, divergent, and scattered light from a sample. Whenever it is desirable to capture the total reflected light from a sample, the integrating sphere is preferred for its light gathering abilities.

The integrating sphere consists of a hollow sphere or hemisphere coated with a highly Lambertian (diffusely reflecting) surface. An important characteristic of an ideal sphere is that the intensity of the reflected energy at any part of the sphere surface is proportional to the total energy entering the sphere. This relationship is ideally independent of the specific point of entry to the sphere and is independent of the incident angle of the energy into the sphere. The main considerations in the use of an integrating sphere include sphere throughput, sphere efficiency, average reflectance at the sphere wall, and sphere error.[10]

The throughput (τ) of an integrating sphere, sometimes referred to as another measure of sphere efficiency, is given by the following expression. Note that τ should be maximized to provide the maximum throughput.

$$\tau = \frac{A_e R_w}{A_S}\left(\frac{1}{\left[1 - R_w\left(\frac{1-A_{T-e}}{A_S}\right)\right]}\right) \tag{2.16}$$

Where A_e = the area of the sphere exit port (to the detector), given by πr^2; R_w = the reflectivity of the sphere wall (e.g., 1.00 = 100%); A_S = the total area of the sphere (e.g., $4\pi r^2$); and A_{T-e} = the total area of all sphere ports minus the area of the exit port(s). Sphere efficiency (E_S) is given by the relationship:

$$E_S = \frac{A_{ep}}{(1-\bar{R})A_S} \tag{2.17}$$

where A_{ep} = the area of the exit port (as πr^2); A_S = the total sphere area (as $4\pi r^2$); and $\bar{R}$ = the average reflectance of the sphere wall including the entrance and exit ports and the sample. The $\bar{R}$ is determined using the relationship:

$$\bar{R} = \frac{R_S A_\sigma + R_w\left(4\pi r^2 - \sum_{i=1}^{n} A_n\right)}{4\pi r^2} \tag{2.18}$$

where R_S = the reflectivity of the sample (e.g., 1.00 = 100%); A_σ = spherical area of the sample port (as πr^2); R_w = the reflectivity of the sphere wall coating (1.00 = 100%); and r = the radius of the sphere. The expression $\sum_{i=1}^{n} A_n$ is the sum of the spherical area of all ports on the sphere, where n = the number of ports; and πr^2 is the spherical area of each port, where r = the radius of the port.

For any integrating sphere, the sphere error (E_S) due to the variance in sphere efficiencies between the sample and the standard when both are measured at the identical port, is given by:

$$E_S = \frac{\varepsilon_S}{\varepsilon_r} - 1 = -\frac{(R_r - R_s)\left(\frac{A_\sigma}{A_S}\right)}{1 - R_w\left(\frac{d}{A_S}\right) - R_\sigma\left(\frac{A_e}{A_S}\right)} \tag{2.19}$$

where ε_s = the sphere efficiency when measuring a sample; ε_r = the sphere efficiency when measuring the reference reflectance standard material; R_r = the reflectance of the reference standard material (e.g. 1.00 = 100%); R_σ = the reflectance of the sample; A_S = the total sphere area; R_w = the reflectance of the sphere wall coating; and $d = A_S - A_{ent} - A_{exit} - A_\sigma$. Where A_{ent}, A_{exit}, and A_σ are the areas of the entrance, exit, and sample ports as πr^2, respectively.

In addition to the above expressions, the total power received at the detector (P_d) as a function of input power (P_i) is given using the expression:[10,13]

$$P_d = \frac{E_S A_d}{\pi\left(\frac{d_{exit}^2}{4} + \delta^2\right)} P_i \tag{2.20}$$

where E_S = the sphere efficiency (as calculated above); A_d = the detector window area; d_{exit} = the exit port diameter; and δ = the distance from the exit port entrance to the detector window. There is a specific section on the "integrating sphere" in this set of volumes and the reader is referred to the index for additional details. A basic design of an integrating sphere for optical measurements is shown in Figure 2.5.

2.1.3 *Total Transmittance/Absorption and Reflectance/Absorption Spectroscopy*

The total transmittance (or transmission) of light through solids or liquids is given by the following Beer-Lambert law relationship. Note: for reflectance spectroscopy,

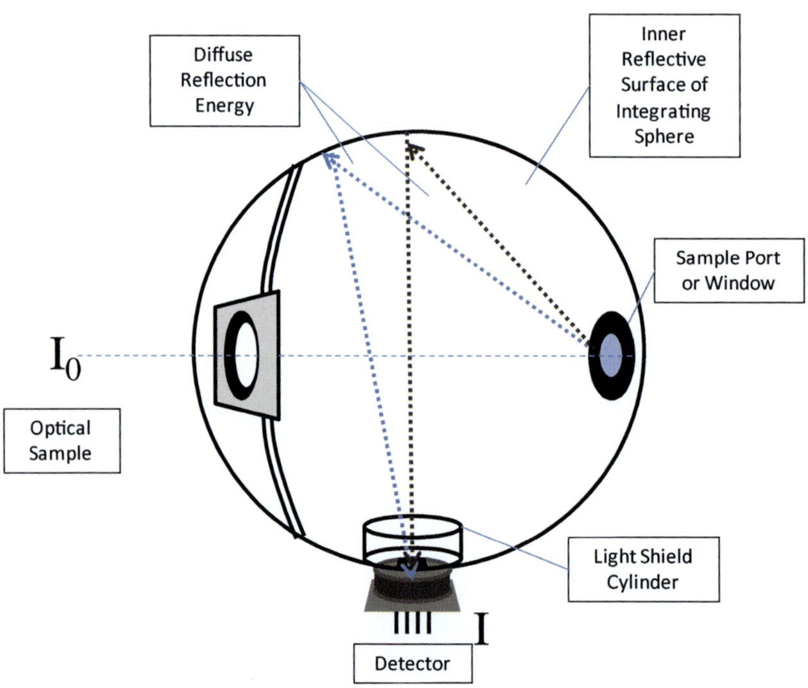

Fig. 2.5 Basic integrating sphere design.

the R term can be substituted for the T term in the following relationships:

$$I_t = I_0 e^{-ecl} \quad \text{or} \quad \ln \frac{I_0}{I_t} = ecl \quad \text{or} \quad I_t = I_0 \times 10^{-ecl} \tag{2.21}$$

where I_t = the intensity of the transmitted or reflected light at a specific wavelength; I_0 = the incident beam intensity at that wavelength; e = the molar extinction coefficient (absorptivity as e, or a or ε) at the wavelength of I_t and I_0; c = the concentration of the absorbing molecules; and l = the thickness of the material. Note: for the above nomenclature; ecl = the total Absorption. This law is most often seen in other equation forms as:

$$\frac{1}{T} = 10^{A} \tag{2.22}$$

and by taking log 10 of each side of the equation we have:

$$\log \frac{1}{T} = A \tag{2.23}$$

Absorbance is related to the concentration of the absorbing molecules as A = ecl. And it follows that:

$$A = -\log_{10}\left(\frac{I}{I_0}\right) = \log_{10}\left(\frac{I_0}{I}\right) \tag{2.24}$$

$$-A = \log_{10}\left(\frac{I}{I_0}\right) \tag{2.25}$$

$$\frac{I}{I_0} = 10^{-A} \Rightarrow \frac{I_0}{I} = \frac{1}{10^{-A}} \tag{2.26}$$

$$\frac{I}{I_0} = T \tag{2.27}$$

$$\%T = T \times 100 \tag{2.28}$$

This relationship between Absorbance (A) and the ratio of I/Io is shown in Figure 2.6.

Transmission spectroscopy is most useful with clear materials, such as thin films, crystalline plates, and clear liquids. It is not particularly useful for opaque or turbid materials, or highly scattering materials.[10,13,20,21]

The measurement should not be used when intermolecular interactions vary with concentration of absorbing molecules; or with solvent type or concentration; or when refractive index changes drastically with concentration. Turbidity will markedly reduce the value of transmission measurements as will significant color changes in the material with dilution or analyte concentration. This straight

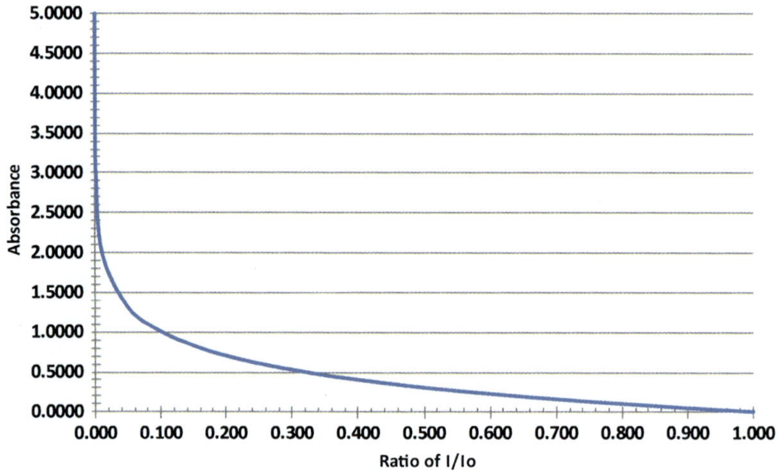

Fig. 2.6 Illustration of relationship between absorbance and I/Io from 0.00001 to 1.0.

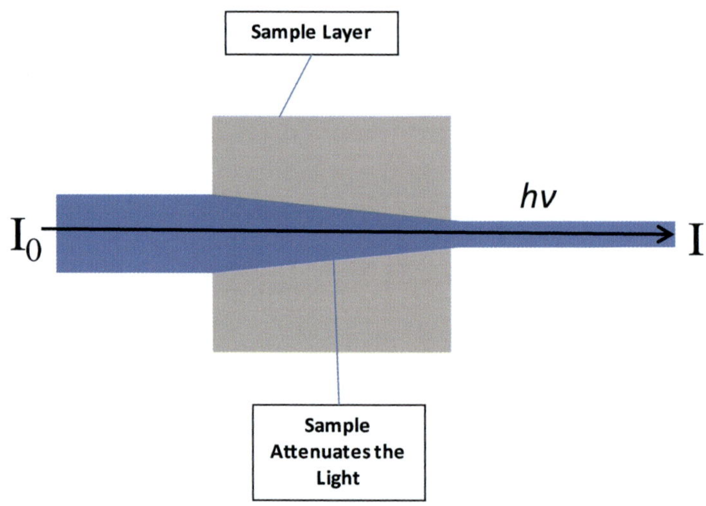

Fig. 2.7 Direct transmittance measurement geometry for solid samples.

through measurement geometry is referred to as direct transmission/transmittance (Figure 2.7).

2.1.3.1 *Diffuse Transmittance*

Diffuse Transmittance (DT) is defined as the total transmitted light passing through (and interacting with) a non-infinite thickness of a diffusely reflecting, or slightly

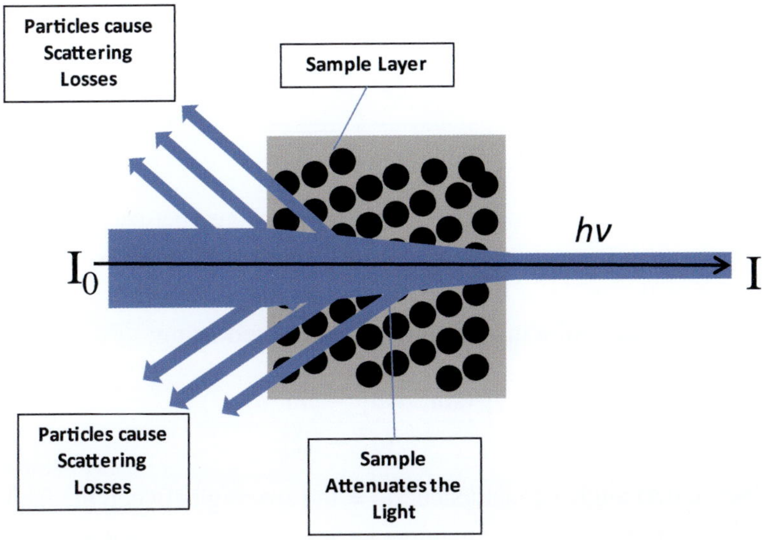

Fig. 2.8 Diffuse transmittance measurement geometry for solid samples.

energy scattering medium; this medium being comprised of multiple diffusely reflective surfaces, like powders or pellets (Figure 2.8).

2.1.3.2 *Specular (Regular) Reflectance*

Specular, external (or regular) reflectance (R_r) is described by the use of the Fresnel Equation as:

$$R_r = \frac{(n_2 - n_1)^2}{(n_2 + n_1)^2} \tag{2.29}$$

where n_2 and n_1 are the refractive indices of two materials, n_2 being the material for which regular reflectance is measured and n_1 being the surrounding medium. For a material in air, this relationship can be expressed as:

$$R_r = \frac{(n_m - 1.0003)^2 + n_m^2 \varepsilon^2}{(n_m + 1.0003)^2 + n_m^2 \varepsilon^2} \tag{2.30}$$

where n_m = the refractive index of the material, and ε = the absorption (or extinction) coefficient of the material.

Applications for external (or regular) reflectance include surface measurements for metals or semiconductors. The technique is used to measure the dielectric function of solids; for characterization of thin films; and to relate the reflectivity of a material surface to its electronic and/or surface structure. Thin film thickness (t, in units of wavelength) is measured using this technique combined with the mathematical relationship:

$$t = \frac{f}{2\sqrt{(n_m^2 - \sin^2\alpha)} \times d} \tag{2.31}$$

where f = the number of complete interference fringes between initial and final fringe interval (peak to peak); d = the wavelength difference for the fringe count interval; and α = the angle of incidence.

The refractive index of a thin film (n_t) is calculated using:

$$n_t = \sqrt{\left[\frac{(\sin^2\alpha_1 d_1^2 - \sin^2\alpha_2 d_2^2)}{(d_1^2 - d_2^2)}\right]} \tag{2.32}$$

where α_1 = the first angle of incidence; d_1 = the spacing between the fringes of the interference pattern created using incident light at angle α_1; α_2 = the second angle of incidence used to create a second interference pattern; and d_2 = the spacing between fringes of the second interference pattern created by light at angle of incidence α_2 (peak to peak).

2.1.4 *Fluorescence*

2.1.4.1 *Introduction*

Photons with energies in the ultraviolet (190 nm to 360 nm or 180 nm to 350 nm for some fluorometers) to the blue-green visible (350 nm to 500 nm) regions of the spectrum will excite an electronic transition in molecules that fluoresce (i.e., fluorophores). Fluorescence is an electronic transition from a ground state to the excited state and with the emission of a photon to return to the ground state. Once excited the molecule relaxes (Stokes shift) to the ground state while emitting a photon within a femtosecond (10^{-15} seconds) to picosecond (10^{-12} seconds) timeframe. The Stokes shift indicates a lower energy of the fluorescence photons than the energy of the excitation photons. The fluorescence typically has a lifetime (or duration) of nanoseconds (10^{-9} seconds) per transition. Figures 2.9 and 2.10 show Jablonski diagrams for the quantum energy transitions for infrared, Rayleigh and Raman scattering, and fluorescence (Figure 2.9); and of fluorescence excitation, radiative energy decay pathways, with photon emission. Note that Ex denotes the energy scale, and T1 is the lowest energy triplet state (Figure 2.10).[22–23]

A fluorescence spectrophotometer normally has an excitation monochromator, that defines the excitation energy, and an emission monochromator that provides a full spectrum of the fluorescence emission. Figure 2.11 demonstrates the schematic

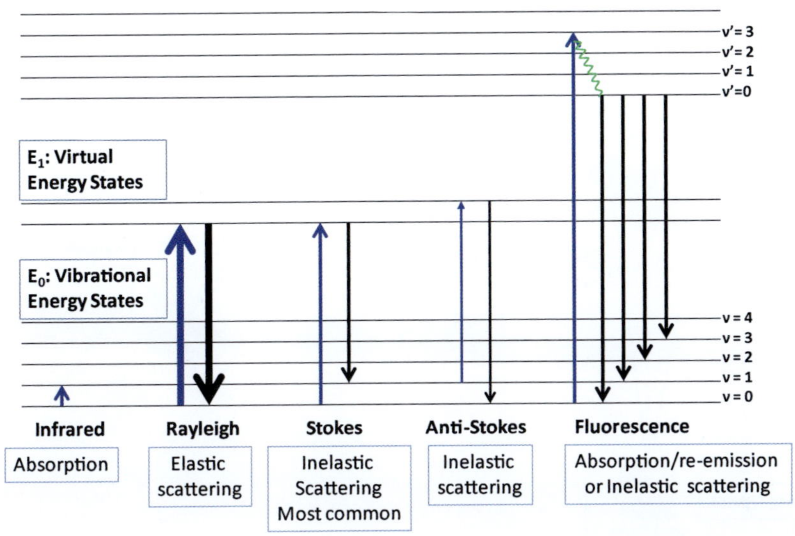

Fig. 2.9 Illustration of Jablonski diagram for the quantum energy transitions for infrared, Rayleigh scattering, Raman scattering, and fluorescence. Thickness of lines indicates the relative energy levels involved for each transition.

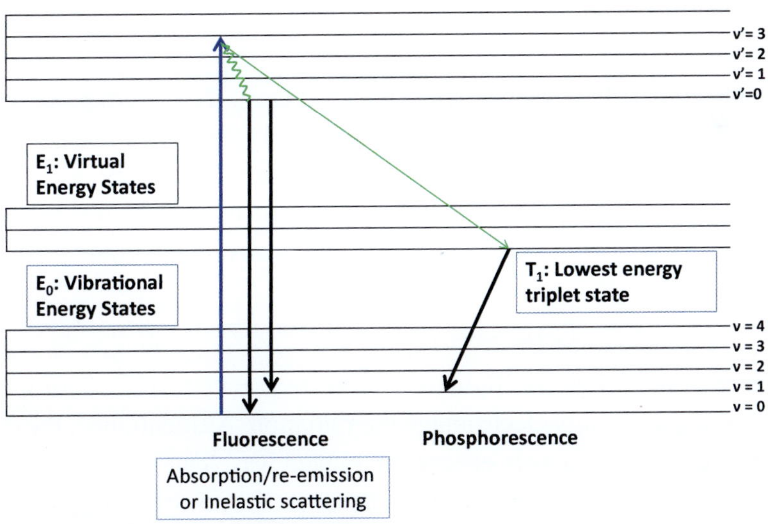

Fig. 2.10 Illustration of Jablonski diagram for the quantum energy transitions for fluorescence and phosphorescence. Shown are the radiative energy level pathways, with photon emission pathways. Note that E_0 and E_1 denote the energy levels, and T1 is the lowest energy triplet state illustrating phosphorescence. Thickness of lines indicates the relative energy levels involved for each transition.

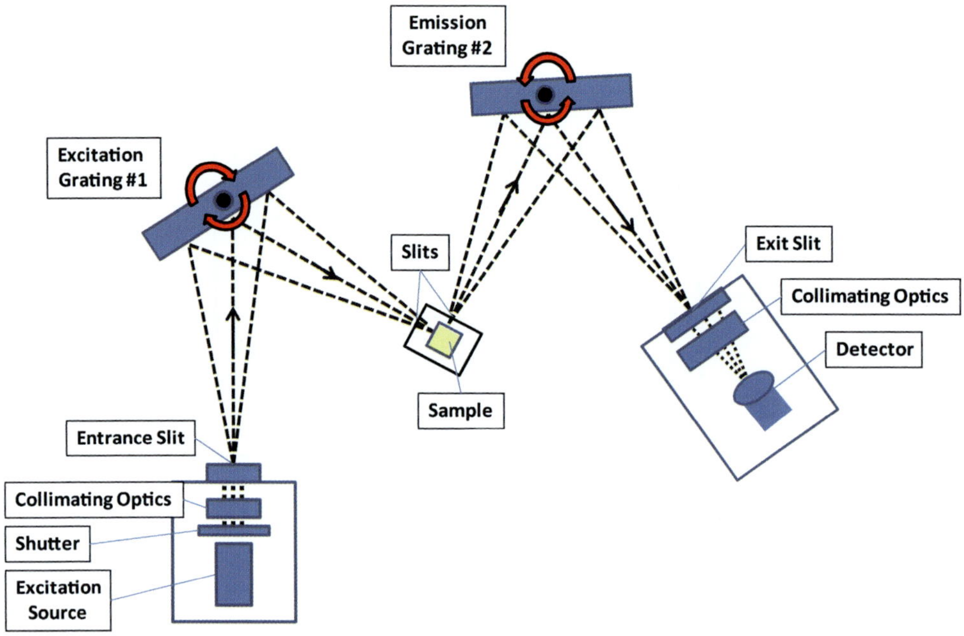

Fig. 2.11 Basic optical schematic illustration of a dual monochromator fluorometer: one excitation monochromator and one emission monochromator. Phosphorescence may also be measured by excitation and then shutting off the source lamp for longer-term phosphorescence emission measurement.

of a fluorometer (or fluorescence spectrophotometer). The overlapping excitation and emission spectra are illustrated in Figure 2.12 for a scanning fluorometer. Note that either the excitation or emission monochromators may be set to a fixed wavelength with the other monochromator set to scan for a given wavelength range. Table 2.1 lists the potential scanning sequences for a research grade fluorometer.

2.1.4.2 *Fluorescence and Phosphorescence*

Fluorescence is the emission of light by a fluorescent material (i.e., fluorophore) that has absorbed higher energy electromagnetic radiation. As mentioned, the emitted or fluorescence energy has lower energy, or longer wavelength, than the excitation radiation. Ultraviolet excitation energy used in fluorescence (typically 180 to 380 nm) provides the most dramatic forms of fluorescence and the emitted energy is often in the visible region (lower energy than the ultraviolet region).

The excitation and emission process of a fluorophore may be written as Equation 2.33.

$$E_0 + h\nu_{ex} \rightarrow E_1 \rightarrow E_0 + h\nu_{em} + heat \tag{2.33}$$

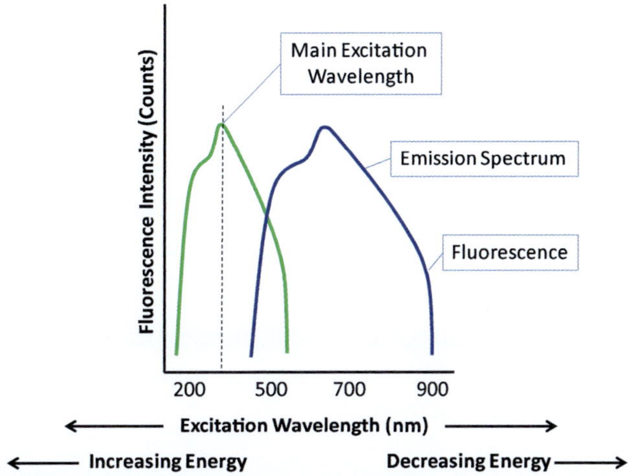

Fig. 2.12 Representation of scanning ranges for a dual monochromator fluorometer. The excitation monochromator may scan from the ultraviolet to the visible range and the emission monochromator may also scan from the ultraviolet to the visible range. The fluorescence is of lower energy (longer wavelength) than the excitation wavelength for typical fluorescence.

Table 2.1 Fluorometer scanning functions.

Excitation wavelength(s)	Emission wavelength(s)	Useful for determining
Scanning	Scanning	Locate EX and EM Wavelengths
Fixed	Scanning	EM Spectrum
Scanning	Fixed	Max EX Wavelength
Fixed	Fixed	Quantitative Analysis

This is symbolic when a lower energy state (E_0) is excited to a higher energy state (E_1) due to excitation energy ($h\nu_{ex}$). The resultant higher energy state remains for approximately a nanosecond (10^{-9} seconds) lifetime and re-emits fluorescence energy in the form of a fluorescent emission photon at lower energy ($h\nu_{em}$). Some energy in the form of heat or infrared is often emitted as well. The term $h\nu$ is used here to represent photon energy as h = Planck's constant and ν = (Gr: nu) = frequency of light. The energy states are E_0, sometimes referred to as S_0 representing the ground state of the fluorophore, and E_1 (or S_1) is the first excited state (referring to the electronic state of the molecule). Heat may also dissipate into a solvent material during the energy state relaxation process (this is referred to as *non-radiative* relaxation energy).

A second type of *non-radiative* relaxation occurs when the excitation energy is dissipated to a triplet state, which then causes a phosphorescence relaxation process. Fluorescence quenching may occur when the energy that would result in fluorescence is absorbed by a second molecule, preventing a fluorescence response or actually decreasing the fluorescence following excitation. Resonance fluorescence may occur when intense excitation energy causes the absorption of 2 photons where the emitted radiation is both at shorter and longer wavelengths than the excitation radiation, this is sometimes referred to as a Raman effect.

2.1.4.3 *Quantum Yield*

The fluorescence quantum yield (Q) gives the efficiency of the fluorescence process. It is defined as the ratio of the number of photons emitted to the number of photons absorbed, as shown in Equation 2.34.

$$Q = \frac{p_{em}}{p_{ab}} \tag{2.34}$$

Where the efficiency is equal to the number of photons emitted (P_{em}) ratioed to the number of photons absorbed (P_{ab}). A highly efficient fluorophore will demonstrate a Q ratio of 0.10 to 0.15; the maximum theoretical Q efficiency is 1.0. Fluorescence quantum yields are measured by comparison to a standard material fluorescence. The quinine salt *quinine sulfate* in a sulfuric acid solution is a common fluorescence standard (see Chapter 8).

2.1.4.4 *Fluorescence Lifetime*

The first-order fluorescence lifetime is the average time the molecule stays in its excited state prior to emitting a lower energy photon. The concentration of the excited state $[E_1]$ at any time is given as Equation (2.35), demonstrating an example of exponential decay.

$$[\mathrm{E1}] = [\mathrm{E1}]_0 e^{-\Gamma t} \tag{2.35}$$

Where [E1] is the concentration of excited state molecules at some time t; $[\mathrm{E1}]_0$ is the initial concentration of molecules at time 0; and Γ is the decay rate equal to the inverse of the fluorescence lifetime. For commonly used fluorescent compounds, typical excited state decay times for photon emissions will range from 0.5 to 20 or more nanoseconds.

2.1.4.5 *Phosphorescence*

Phosphorescence is a form of photoluminescence with a slower time scale of re-emission than fluorescence. Phosphorescence is associated with "forbidden" energy state transitions that occur very slowly. This process may emit radiation at a low energy rate for several hours after absorption of excitation photons. It may be measured using most spectrofluorometers (Figure 2.11).

References

1. J.J. Workman and A.W. Springsteen, (Eds.), *Applied Spectroscopy: A Compact Reference for Practitioners* (Academic Press, Boston 1998).
2. J.W. Blaker, *Optics II — Physical and Quantum Optics* (Barnes and Noble, New York, 1970).
3. R.W. Ditchburn, *Light*, 2nd ed. (Blackie, London 1965).
4. M. Fogiel, (Ed.), *The Optics Problem Solver* (Research & Education Association, 1981).
5. J.W. Goodman, *Statistical Optics* (Wiley, New York, 1985).
6. F.A. Jenkins and H.E. White, *Fundamentals of Optics* (McGraw-Hill, New York, 1957).
7. C.S. Johnson, Jr. and L.G. Pedersen, *Problems and Solutions in Quantum Chemistry and Physics* (Dover, New York, 1986).
8. E. Mach, *The Principals of Physical Optics* (Dover, New York, 1926).
9. B. Crosignani, P. DiPorto, and M. Bertolotti, *Statistical Properties of Scattered Light* (Academic Press, New York, 1975).
10. G. Kortum, *Reflectance Spectroscopy* (Springer-Verlag, New York, 1969).
11. V.P. Kubelka and F. Munk, Ein beitrag zur optic der fabanstriche, *Z. Tech Physik* **12** (1931), 593–601.
12. A. Schuster, Radiation Through a Foggy Atmosphere, *Astrophysical Journal* **21** (1905), 1–22.
13. W.W. Wendlandt and H.G. Hecht, *Reflectance Spectroscopy* (Wiley, New York, 1966).
14. R.D. Braun, *Introduction to Instrumental Analysis* (McGraw-Hill, New York, 1987).
15. G.L. Clark, (Ed.), *The Encyclopedia of Spectroscopy* (Reinhold, New York, 1960).
16. C.J. Ball, An Introduction to the Theory of Diffraction (Permagon, Oxford, 1971).
17. J.B. Cohen, *Diffraction Methods in Material Science* (Macmillan, New York 1966).
18. J. Guild, *Diffraction Gratings as Measurement Scales* (Oxford University Press, London, 1960).
19. W.R. Ware, Transient Luminescence Measurements, *In Vol. 1 — Creation and Detection of the Excited State*, A.A. Lamola (Ed.), (Marcel Dekker, New York 1971).
20. G.A. Somorjai, (Ed.), *The Structure and Chemistry of Solid Surfaces* (Wiley, New York, 1969).
21. D.P. Woodruff and T. A. Delchar, *Modern Techniques of Surface Science*, (Cambridge University Press, Cambridge, 1986).
22. P.T.C. So, C.Y. Dong, *Fluorescence Spectrophotometry, In Encyclopedia of Life Sciences*, 1st edn. (Macmillan Publishers Ltd, Nature Publishing Group, 2002).
23. J.R. Lakowicz, *Principles of Fluorescence Spectroscopy,* 3rd edn. (Springer Science & Business Media, 2007).

Chapter 3

Ultraviolet Spectroscopic Instrumentation

Contents

3.1 Instrumentation for Measuring Ultraviolet Spectra

Spectrophotometers used within the laboratory for ultraviolet measurements may function using a variety of optical and mechanical design configurations. Spectrophotometers have a technical definition as a device which provides the ratio, or a function of the ratio, of the radiant power of an incident energy beam for both sample and reference measurements (two beams) as a function of spectral position (i.e., wavelength or frequency). The sample and reference beams may be separated in time, space, or both. This definition is paraphrased from the ASTM E131 definition for a spectrophotometer.[1]

The measurement for any ultraviolet spectrophotometer is given by the ratio of the measurement of the sample at a specific wavelength or set of wavelengths (as I) divided by the measurement of a reference material at the same wavelength or set of wavelengths (I_0). If the measurements are taken with a transmittance geometry, then this ratio is equal to T (i.e., the transmittance) with a value of from 0 to 1.0. If the measurements are taken with a reflectance geometry, then this ratio is equal to R (i.e., the reflectance) also with a value of from 0 to 1.0. For transmittance reference measurements a blank sample holder or ambient air is generally used. For reflectance reference measurements a highly reflective sample is used. This reflective reference is generally a ceramic, or Spectralon™ sample each having a surface that is 95 to 99 percent reflective for each wavelength region measured.

A common assumption in spectrophotometric measurements is that Beer's law relationship (repeated as core information in different forms and contexts throughout this set of volumes, except for fluorescence and Raman spectroscopy) applies for a change in instrument response due to a change in concentration of analyte molecules present within a sample specimen. *Beer's law* assumes that the measured transmittance, or measured reflectance, of a sample within an incident beam is equivalent to 10 exponent the negative product of the absorptivity (in Liter mol^{-1} cm^{-1}), times the concentration of a molecule in solution (in mol $Liter^{-1}$), times the pathlength (in cm) of the sample in solution (Equation 3.3). However changes in sample and instrument temperature, atmospheric pressure within the instrument, and interferences caused by overlapping absorption bands will create deviations from the ideal case. For most analytical situations Beer's law relationship holds true enough for useful quantitative measurements. Note that temperature changes of optics, including lenses and cut-on or cut-off filter profiles may cause variations in spectra with temperature and small mechanical differences in alignment. These minute changes may be modeled and eliminated when properly understood. The use of special materials, such as carbon fiber mechanical parts may also reduce temperature effects.

The meaning of Beer's law for ultraviolet spectroscopic measurement follow a functional description such that the type and frequency of absorption determines the frequency and appearance of the ultraviolet spectrum. The amplitude of the absorption at any particular wavelength or wavenumber is determined by its absorptivity of the molecules, and the number of molecules encountered, within the beam path of the measuring instrument. It is assumed that a change in spectral response is related to a concentration as described by the Bouguer, Lambert, and Beer relationship most often termed Beer's law. The Beer's law relationship is described as the absorbance (A, Au, or signal strength) of an analyte being measured using a spectrophotometer and is equivalent to the product of the absorptivity (ε) of a specific type of molecular vibration; the concentration (c) of the molecules in the measurement beam; and the pathlength (l) of the sample holder within the measurement beam. This relationship between measured spectral signal and concentration of a molecule is most often expressed as:

$$\mathrm{A} = \varepsilon \mathrm{cl} \tag{3.1}$$

where ε is the molar absorptivity (referred to as molar extinction coefficient by earlier physicists) in units of Liter $\cdot$ Mole^{-1} $\cdot$ cm^{-1}; c is the concentration of molecules in the spectrometer beam in units of Mole $\cdot$ Liter^{-1} (as moles per volume, or weight percent per volume); and pathlength is the thickness in units of cm of the measured sample at a specific concentration. The absorptivity for any specific molecule type is calculated by careful measurements of the absorbance of a compound, generally diluted in a suitable organic solvent, and by applying the relationship:

$$\varepsilon = \frac{A}{cl} \tag{3.2}$$

Note that for transmittance (where T = 0.0 to 1.0) and percent transmittance (where %T = 0 to 100.0) spectroscopy, a more complete delineation of the relationships between the various terms is contained in an expression such as:

$$T = \frac{I}{I_0} = 10^{-\varepsilon cl} \Rightarrow Abs. = A = -\log_{10}\left(\frac{I}{I_0}\right) = -\log_{10} T = \varepsilon cl \tag{3.3}$$

Here the symbols I and I_0 represent the attenuated energy detected after sample interaction and the initial energy incident to the sample, respectively. For reflectance (where R = 0.0 to 1.0) and percent reflectance (where %R = 0.0 to 100.0) spectroscopy the various relationships are expressed as:

$$R = \frac{I}{I_0} = 10^{-\varepsilon cl} \Rightarrow Abs. = A = -\log_{10}\left(\frac{I}{I_0}\right) = -\log_{10} R = \varepsilon cl \qquad (3.4)$$

Noting the relationship exists where the change in intensity (I) of the transmitted or reflected light from a sample is a function of the change in pathlength (l) of the sample as expressed by the absorptivity (ε) of a specific analyte (or molecular substance) and its concentration (c) by:

$$-\frac{\delta \mathrm{I}}{\delta \mathrm{l}} = \varepsilon \mathrm{cI} \cdot (\ln 10) \qquad (3.5)$$

Modern absorption based spectrophotometers utilize these assumptions for making spectroscopic measurement and generally display spectroscopic data as transmission (T), reflection (R), and absorbance A (y-axis, or ordinate axis) versus wavelength (nanometers or microns) or wavenumbers (cm^{-1}) (as x-axis, or abscissa axis).

With a basic knowledge of Beer's law one is able to calculate the absorptivity and the other parameters of Absorbance, pathlength, and concentration for various sample measurements. The relationship between Absorbance and Transmittance or Reflectance is given in Tables 3.1 and 3.2.

The use of the formula:

$$\mathrm{A} = \varepsilon \mathrm{cl} \qquad (3.6)$$

allows the simple calculation of the absorptivity as:

$$\varepsilon = \frac{A}{lc} \text{ in units of } \mathrm{L} \cdot \mathrm{mol}^{-1} \cdot \mathrm{cm}^{-1} \qquad (3.7)$$

Table 3.1 Showing the relationships between %T and %R, T and R, and A.

%Transmittance or %Reflectance	Transmittance or Reflectance	Absorbance
100.0	1.0	0.0
10.0	0.1	1.0
1.0	0.01	2.0
0.1	0.001	3.0
0.01	0.0001	4.0
0.001	0.00001	5.0
0.0001	0.000001	6.0

Table 3.2 Table showing the relationship of Absorbance unit (Au) values to Reflectance and Transmittance values and to %R and %T.

Absorbance	Reflectance/ Transmittance	% Reflectance/ % Transmittance
3.5	0.00032	0.032
3.0	0.00100	0.100
2.0	0.01000	1.000
1.5	0.03162	3.162
1.0	0.10000	10.00
0.5	0.31623	31.62
0.1	0.79433	79.43
0.01	0.97724	97.72

And one can readily calculate the reflectance or transmittance of any measurement given the absorbance of that measurement as follows. For conversion of Absorbance to Transmittance we use:

$$A = -\log_{10}\left(\frac{I}{I_0}\right) = \log_{10}\left(\frac{I_0}{I}\right) \tag{3.8}$$

$$-A = \log_{10}\left(\frac{I}{I_0}\right) \tag{3.9}$$

$$\frac{I}{I_0} = 10^{-A} \Rightarrow \frac{I_0}{I} = \frac{1}{10^{-A}} \tag{3.10}$$

$$\frac{I}{I_0} = 10^{-A} = T \tag{3.11}$$

$$\%\mathrm{T} = \mathrm{T} \times 100 \tag{3.12}$$

To convert Absorbance to Reflectance we use:

$$A = -\log_{10}\left(\frac{I}{I_0}\right) = \log_{10}\left(\frac{I_0}{I}\right) \tag{3.13}$$

$$-A = \log_{10}\left(\frac{I}{I_0}\right) \tag{3.14}$$

$$\frac{I}{I_0} = 10^{-A} \Rightarrow \frac{I_0}{I} = \frac{1}{10^{-A}} \tag{3.15}$$

$$\frac{I}{I_0} = 10^{-A} = R \tag{3.16}$$

$$\%\mathrm{R} = \mathrm{R} \times 100 \tag{3.17}$$

The goal in the design of laboratory spectrophotometers is to maximize the energy (or radiant power) from a light source through the optical components and onto the sample and then the detector. The optical throughput for a spectrometer is dependent upon multiple factors, such as the light source illumination area, the apertures present within the light path (such as entrance slits and J-stop), lens transmittance and mirror reflectance losses, the exit aperture(s), and the detector efficiency and linearity.[2–3]

Several terms are useful in any discussion of spectrometry or spectrophotometry, these include *selectivity* or the specific sensor response to the *component or analyte of interest; sensitivity* or the *quantifiable level of response* from a sensor with respect to the concentration of a specified component of interest; and *detection limit* or the smallest concentration difference that can be detected above the background noise level of the instrument. A quick estimate of the detection limit is approximately three times the background noise level, and estimating the concentration for this signal level using a calibration curve. Since random noise is normally distributed, the detection limit for an analyte would be represented by a signal amplitude greater than a 99.7% peak-to-peak amplitude (i.e., three times the noise level) as representative of a value greater than three standard deviations.

This noise amplitude, when using multivariate calibration models, is generally approximated by the standard error of cross validation (SECV) or standard error of prediction/performance (SEP). The SECV or SEP may be used to estimate the relative error (i.e., standard deviation of noise) expected when predicting results. The standard error of estimate (SEE) is not useful as a predictor of actual error in analysis. (See Volumes 3 and 5, Chapter 6 and the index for more information on this).

3.1.1 *Types of Spectrophotometers*

3.1.1.1 *Discrete Photometers*

Discrete photometers consist of an irradiance source, discrete interference filters, a sample compartment, and detector; along with appropriate electronics for signal amplification and stabilization of detector signals. The optical configuration for a generic photometer including an electronics overview is shown in Figure 3.1.

3.1.1.2 *Single-Beam*

Single-beam spectrophotometers have a single optical channel which is configured to measure either sample, or reference channel, but not both simultaneously. The resultant spectrum is the ratio of the transmission (or reflectance) spectra from

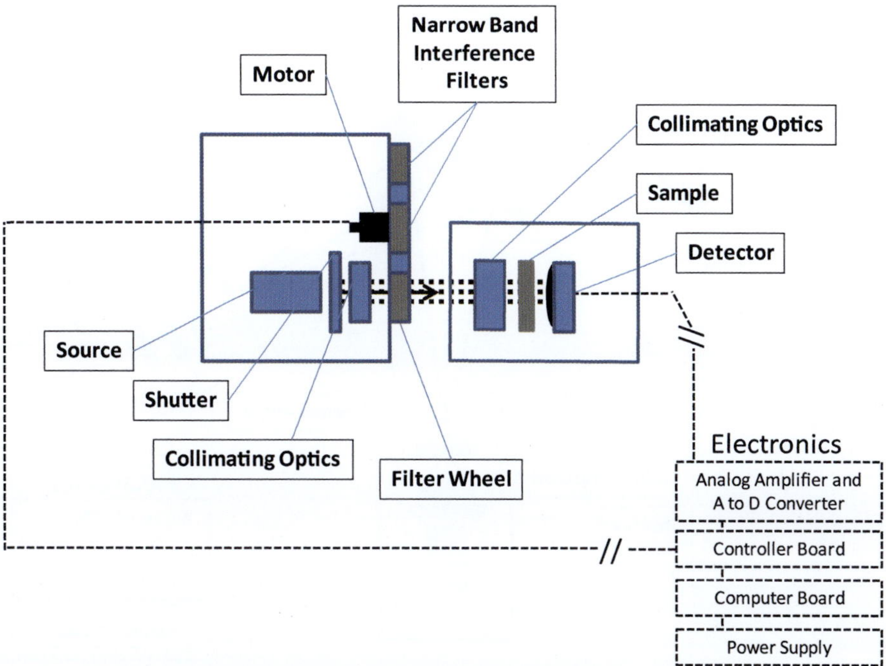

Fig. 3.1 Interference filter photometer instrument configuration.

both sample and reference measurements. The final transmission (in T units) or reflectance (in R units) for a spectrum measured using this device is given as Equation 3.18.

$$\text{Reflectance or Transmittance as } \left(\frac{I}{I_0}\right) = \left(\frac{\text{Sample} - \text{DarkSignal}}{\text{Reference} - \text{DarkSignal}}\right) \quad (3.18)$$

This ratio is reported as 0 to 1.0, or as percent from 0 to 100.0%. If the instrument has any instabilities (either optical, mechanical, or electronic), the time delay between sample and reference measurements must be as short as possible. Single-beam instruments must either be more inherently stable or alternate between sample and reference measurements at a frequency rapid enough so as to negate the rate of change of the spectrometer. Designs for single and double monochromator spectrophotometers, as well as diode array design spectrophotometers are shown in Figures 3.2 through 3.4. These designs can be configured as single- or double-beam. Figures 3.2a and 3.2b demonstrate the single monochromator design with single- or double-beam configurations, respectively.

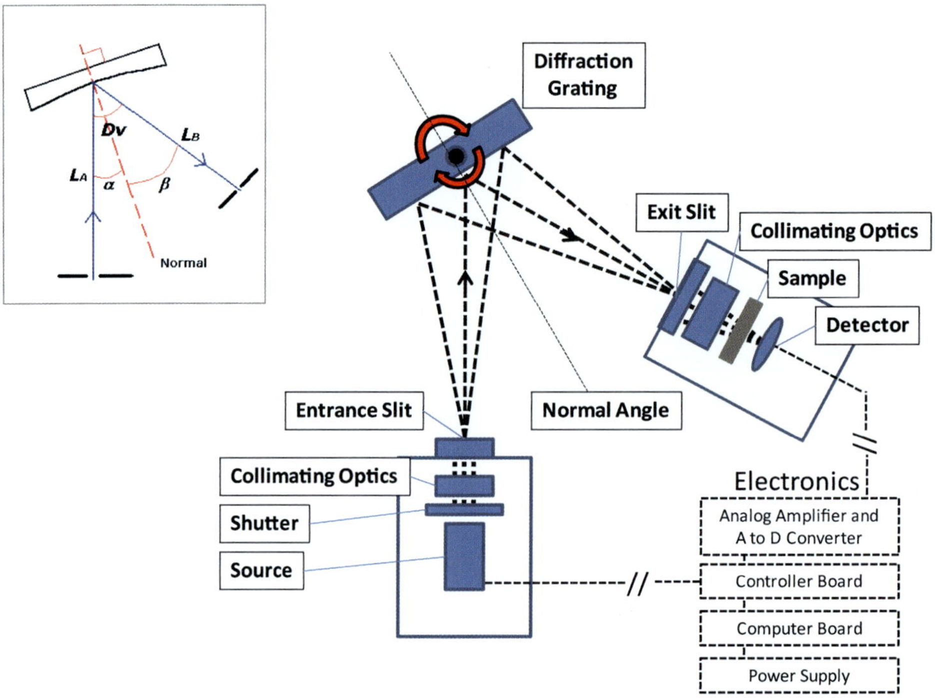

Fig. 3.2a Single monochromator system (dispersive) instrument configuration. Single beam configuration.

If a single internal reference is used for spectral collection, then the absorbance spectrum with respect to wavelength is computed as:

$$A = -\log_{10}\left(\frac{I}{I_0}\right) = -\log_{10}\left[\left(\frac{\mathrm{S}-\mathrm{D_S}}{\mathrm{R_I}-\mathrm{D_{R_I}}}\right)\right] \tag{3.19}$$

If an external standard material is used to calibrate the internal reference material, then the absorbance spectrum with respect to wavelength is computed as:

$$A = -\log_{10}\left(\frac{I}{I_0}\right) = -\log_{10}\left[\left(\frac{\mathrm{S}-\mathrm{D_S}}{\mathrm{R}_I+\mathrm{R}_\Delta-\mathrm{D_{RI}}}\right)\right] \tag{3.20}$$

where:

$$\mathrm{R}_\Delta = (\mathrm{R_E}-\mathrm{D_{RE}})/(\mathrm{R_I}-\mathrm{D_{RI}}) \tag{3.21}$$

and where S is the sample measurement; $\mathrm{D_S}$ is the Dark measurement for the Sample; $\mathrm{R_I}$ is the internal Reference Measurement; $\mathrm{D_{R1}}$ is the Dark Measurement for the Internal Reference Sample; $\mathrm{R_E}$ is the measurement of the external standard reference material for calibration; $\mathrm{D_{RE}}$ is the Dark Measurement for the external standard reference material; and R_Δ is the difference between the dark corrected $\mathrm{R_E}$

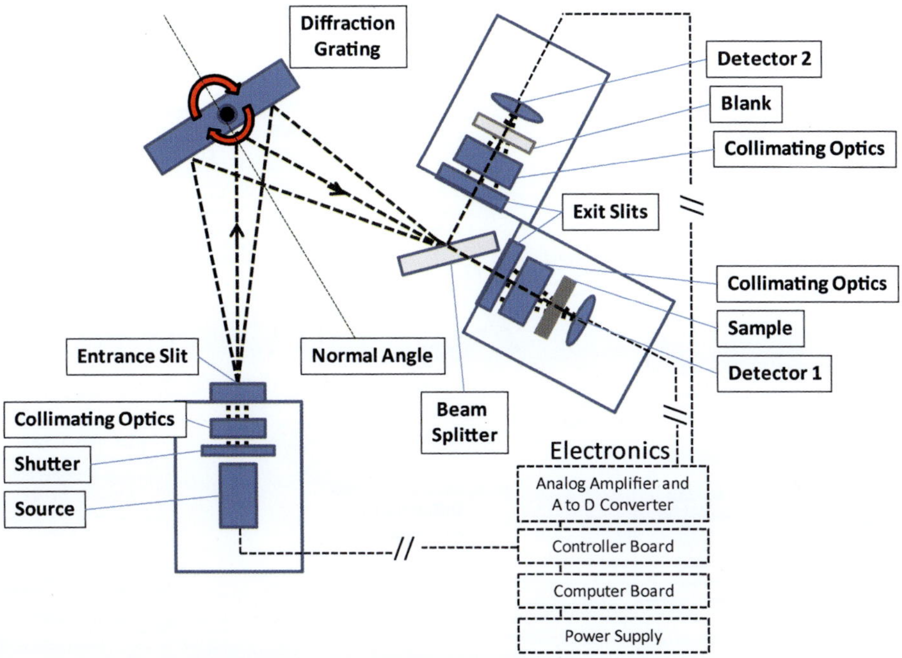

Fig. 3.2b Single monochromator system (dispersive) instrument configuration. Double (dual) beam configuration.

and R_I. The final spectrum is simplified to the ratio of (Sample-Dark) divided by (Internal Reference corrected by External Reference minus Dark). Note that Dark is a measurement where no energy from the source is allowed to the detector (i.e., the source energy is blocked).

The Dark measurement represents the dark current changes in the instrument during a measurement that relate to electronic noise where the detection electronics, computational electronics, and control electronics perform basic measurement functions with no imputed energy reaching the detector (i.e., the source is turned off or blocked using a shutter closure during the Dark measurement).

3.1.1.3 *Double-Beam*

A double-beam spectrometer/spectrophotometer consists of both sample and reference channels and measures both channels simultaneously. The separation of the light beam from the source is accomplished using a fixed beam splitter or alternatively a rotating partial mirror. For the beam splitter, 50% of the source emitted energy is divided to both sample and reference channels and these are measured simultaneously and the ratio computed. The use of the dual-beam concept compensates for instrument instabilities inherent to all spectrophotometers. The

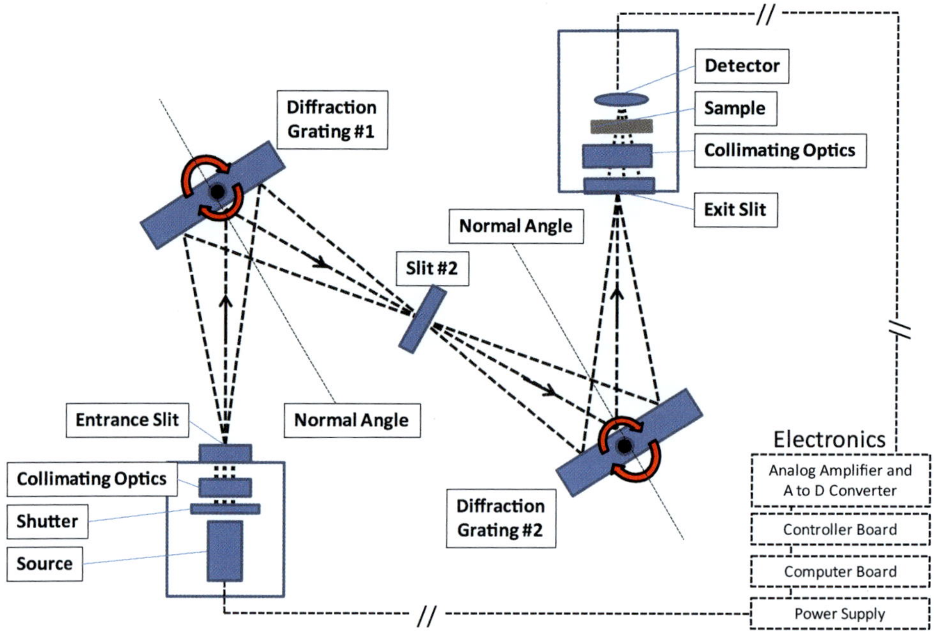

Fig. 3.3 Double monochromator system (dispersive) instrument configuration.

resultant spectrum is the ratio of the sample and reference channels in transmission or reflectance, as in the case of single beam instruments. In practice, the final spectrum from this device is given in the form of Reflectance or Transmittance measurements as Equation 3.22.

$$\left(\frac{I}{I_0}\right) = \left(\frac{\text{Sample} - \text{DarkSignal}}{\text{Reference} - \text{DarkSignal}}\right) = \left(\frac{\mathrm{I} - \delta}{\mathrm{I}_0 - \delta}\right) \tag{3.22}$$

3.1.1.4 *Fluorometers*

A research grade fluorometer consists of a dual monochromator system with an excitation and emission scanning capability. Either the excitation or emission monochromator may be set to a fixed wavelength, or set for a scanning measurement over a fixed wavelength range. A variety of experiments may be completed using the fixed wavelength and scan wavelength features of the fluorometer. Figure 3.5 illustrates the schematic of the dual monochromator fluorometer design. Figures 3.6a through 3.6d illustrate potential scanning experiments available using a dual monochromator fluorometer/spectrofluorometer.

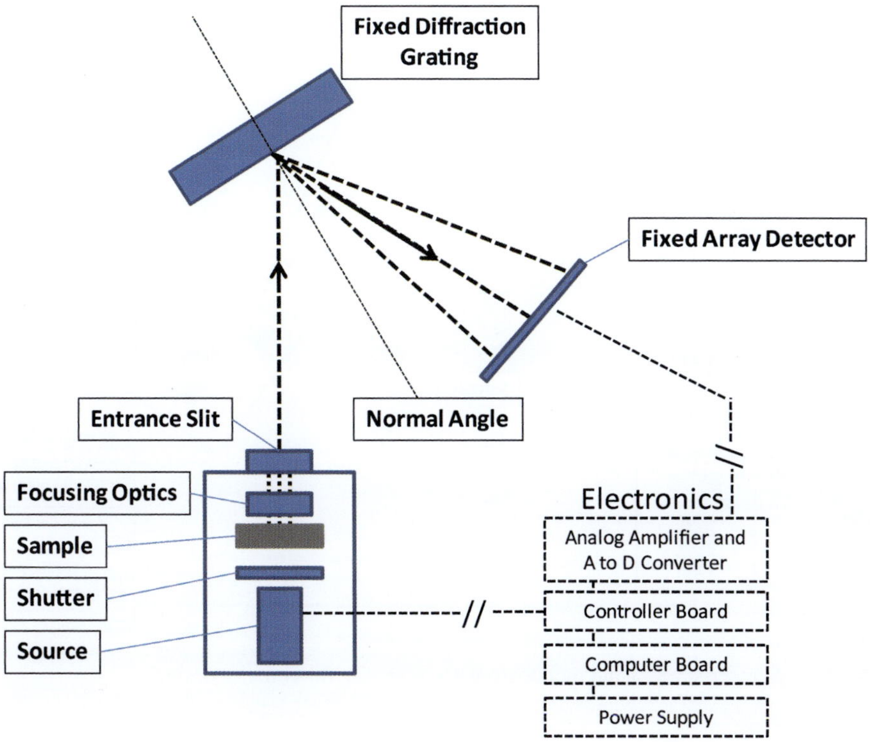

Fig. 3.4 Fixed array detector spectrophotometer (dispersive) instrument configuration.

3.1.2 *Spectrometer Components*

3.1.2.1 *Light Sources*

All materials emit electromagnetic radiation, the nature of which depends upon the material and its temperature. Gases or vapors of atoms at less than atmospheric pressure emit energy/radiation at discrete wavelengths when an electric current is passed through them. Excited gases containing molecular forms emit spectra consisting of multiple lines very close together which comprise emission bands. Solid materials emit or radiate continuous spectra across all frequencies. Comprehensive lists of energy sources applied for optical spectrophotometers are shown in Tables 3.3 and 3.4.

The following principles define the physics of spectrometer energy sources used for laboratory and process instrumentation. Kirchhoff's law states that for a thermal radiator at constant temperature and in thermal equilibrium, the emissivity (e) at any given frequency is equal to the absorptance (a) for radiation from the same direction

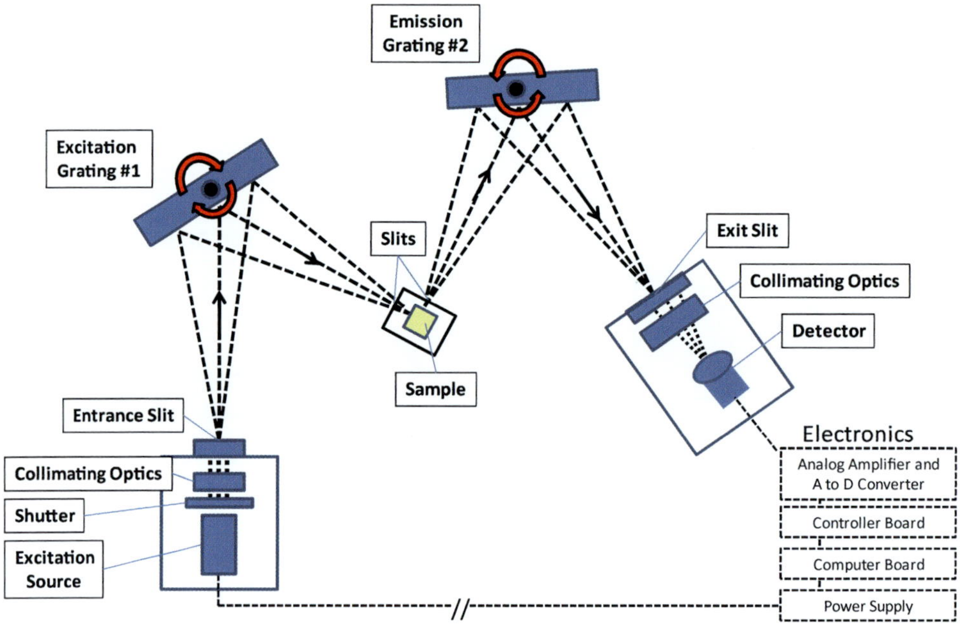

Fig. 3.5 Dual monochromator scanning fluorometer design.

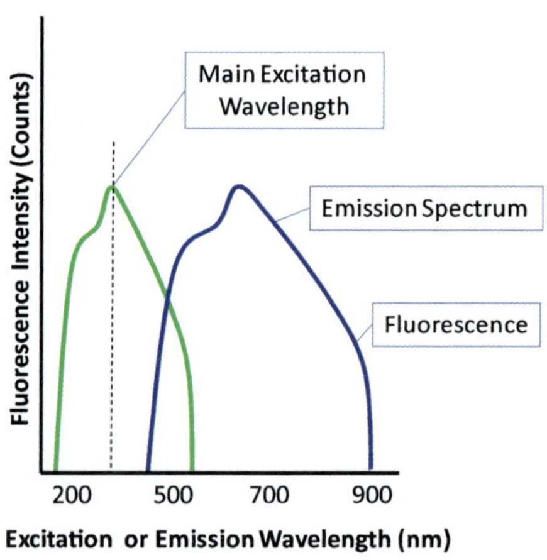

Fig. 3.6a Dual monochromator scanning fluorometer with scanning wavelength excitation and scanning emission experiment.

so that $e = a$, where *absorptance* is defined as the ratio of energy absorbed by the surface of a material ratioed to the total energy striking the surface.

An ideal surface absorbing all the energy that strikes its surface could be termed an ideal *black body* (no energy is reflected or emitted, thus it appears perfectly pure

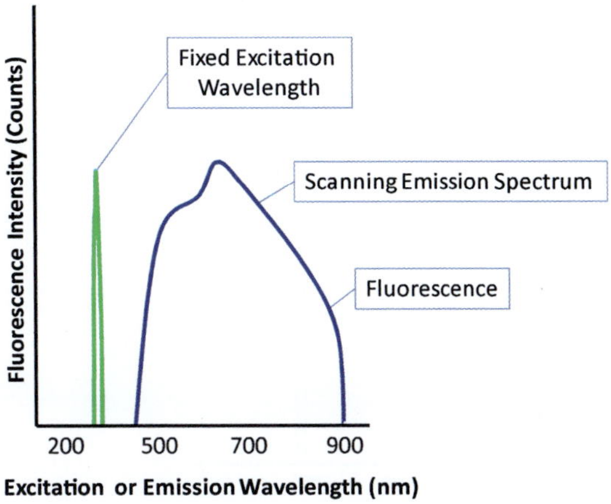

Fig. 3.6b Dual monochromator scanning fluorometer with fixed wavelength excitation and scanning emission experiment.

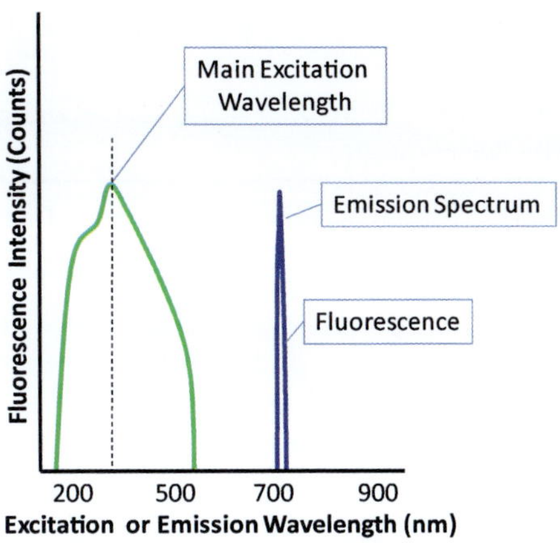

Fig. 3.6c Dual monochromator scanning fluorometer with scanning wavelength excitation and fixed emission experiment.

black, or fully absorbing, at all frequencies). It also stands from Kirchhoff's law of thermal radiation that an ideal black body would also be an ideal emitter of radiation. By definition the emissivity (ε) for any near black body source is defined as the ratio of the actual emitted radiance (ρ) to the ideal emitted radiance of a perfect black body (ρ_β) at the same temperature and frequency (or wavelength). The following ratio illustrates this principle.

$$\varepsilon = \frac{\rho}{\rho_\beta} \tag{3.23}$$

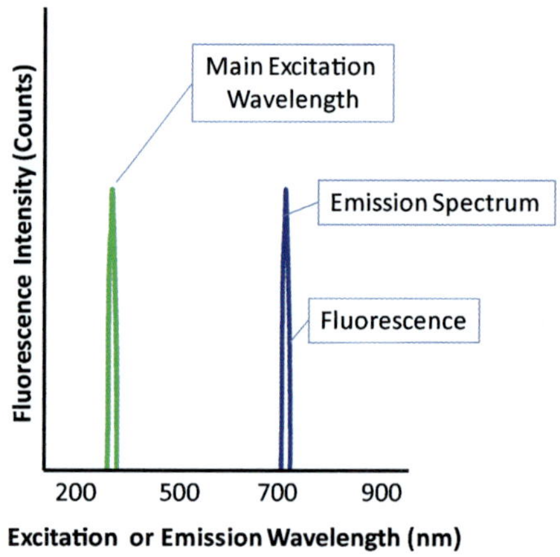

Fig. 3.6d Dual monochromator scanning fluorometer with fixed wavelength excitation and fixed emission experiment.

Table 3.3 Various spectrometer energy sources and effective emissivity ranges (in μm).

Source emissivity ranges (μm)	Start (μm)	End (μm)
Quartz tungsten halogen	0.22	2.7
Glass tungsten halogen	0.25	2.25
DC deuterium lamp	0.185	3.75
Pulsed xenon arc lamp	0.18	2.5
DC arc lamp	0.2	2.5
Mercury lamp	0.3	100
Visible Range	0.36	0.78

Table 3.4 Most common spectrometer energy sources and effective emissivity ranges.

Source	Useful emission range (nanometers)	Useful emission range (cm^{-1})
Quartz Tungsten-halogen mono-filament Lamp	220 – 2700	45,455 – 3,704
Pulsed Xenon Arc Lamp	180 – 2500	55,556 – 4,000
DC Arc Lamp	200 – 2500	50,000 – 4,000
Mercury lamp	300 – 100,000	33,333 – 100

The spectral radiance as measured (ρ) or ideal (ρ_β) is defined as the radiant power (or photon flux density) emitted per unit source area per unit solid angle (typically denoted in W/sr/m^2 (i.e., watts per steradian per square meter). See Equations 3.25 and 3.26 for more definition of the $\rho_{\beta \text{ term}}$.

Max Planck derived the principles of quantum mechanics and expressed this using the concept of light quanta as discrete energy packets that are transferred with an energy (E) proportional to the frequency of the electromagnetic radiation involved. Thus, *Planck's law* has become one of the most widely known concepts in the physics of electromagnetic radiation. This expression of Planck's constant is given as follows:

$$\mathrm{E} = \mathrm{h}\nu \tag{3.24}$$

where E is the energy content of each quantum of light/energy; and Planck's constant (h) with uncertainty, as the last two digits in parentheses, is 6.62606957(29) $\times$ 10^{-34} J· sec.; and frequency (nu $= \nu$) is in sec.$^{-1}$. From Planck's work, the Planck's radiation formula is used to calculate the radiance emitted from an ideal black body (ρ_β) at a given wavelength (λ), emissivity (e), and temperature (T) as follows:

$$\rho_\beta = 3.745 \times 10^4 \times \lambda^{-5} \left(\frac{1.4388 \times 10^4}{e^{\lambda T} - 1} \right) \tag{3.25}$$

where ρ_β is in units of W/sr/m^2 (i.e., watts per steradian per square meter); or W/μm/cm^2 (i.e., watts per micron per square centimeter) denoting spectral radiance.

Further from Planck's radiation formula, other important relationships describing a black body spectrum can be derived. The Stefan-Boltzmann law gives the total radiance emitted (ρ_β) from a black body as a well-defined function of the black body temperature. This relationship is summarized as follows:

$$\rho_\beta = \mathrm{S} \times \mathrm{T}^4 \tag{3.26}$$

where S is the Stefan's constant as 5.67 $\times$ 10^{-8} W/(m^2 °K^4).

The Wien's displacement law demonstrates the shift to shorter wavelengths of the maximum peak position for a black body spectrum as a function of temperature. This means that the higher the temperature of an energy source as a black body radiator the lower is the peak emissive energy. This is given by the following relationship:

$$\lambda_{\max} = \frac{2898 \mu m \cdot {}^\circ K}{T} \tag{3.27}$$

where $\lambda_{\max}$ is the peak maximum position from the black body spectrum in micrometers/microns (μm), and T is in °Kelvin. [Note: °K =° C + 273.]

To express Lamp output in terms of Amps, Volts and Watts the formula used is:

$$\mathrm{A} \times \mathrm{V} = \mathrm{W} \tag{3.28}$$

$$\text{Amps} \times \text{Volts} = \text{Watts} \tag{3.29}$$

These are derived from the relationships,
Ohm's law:

$$\mathrm{V} = \mathrm{IR} \quad \text{or} \quad \mathrm{W} = \mathrm{A}\Omega \tag{3.30}$$

Electrical Power Relationship:

$$\mathrm{P} = \mathrm{IV} \quad \text{or} \quad \mathrm{W} = \mathrm{AV} \tag{3.31}$$

Noting that V = Voltage (volts or V); I = Current (amps or amperes or A), R = Resistance (Ohm's or Ω), and P = Power (watts or W).

3.1.2.2 *Detectors (Performance and Sensitivity)*

There are two basic types of photon (photo) detectors, these being photoemissive and solid state. The photoemissive type is generally represented by the photomultiplier tube detectors; whereas the solid state type detectors are represented by photodiode detectors, pyroelectric detectors, and some photo detecting infrared detectors (e.g., photoconductive and photovoltaic), see Table 3.5.

In defining the physics of detector device performance or sensitivity, two basic terms deserve explanation. First the term specific detectivity or D-star (D*) is essential. This D* value is defined as the detectivity of a radiation detector as a function of the square root of the product of the active detector element area (A in cm^2) and the bandwidth (ω) in units of cycles per second (Hertz, Hz); divided by the noise equivalent power (NEP) in Watts for the detector element. This is summarized by the following equation:

$$\mathrm{D}^* = \frac{\sqrt{A \cdot \omega}}{NEP} = \mathrm{cm} \cdot \mathrm{Hz}^{1/2}\mathrm{W}^{-1} \tag{3.32}$$

The D* is reported in units of $\mathrm{cm} \cdot \mathrm{Hz}^{1/2}\mathrm{W}^{-1}$ units and is used to compare a photodetector responsivity as a function of wavelength. These are referred to as *Jones* units, for the namesake of R. Clark Jones.

For photo-detectors, detector responsivity (R) or input-output gain, is also used to express detector responsivity in units of Amperes per Watt (A/W) as a function of the frequency or wavelength of light incident to the detector. This is a measure

Table 3.5 Most common spectrometer detectors and effective detectivity ranges (in μm).

Detector sensitivity ranges (μm)	Start (μm)	End (μm)	Difference (μm)
Silicon (Si)	0.3	1.1	0.8000
Silicon Photodiode Detectors	0.19	1.1	0.9100
Photomultipler Tube (PMT), S1 Type Photocathode	0.3	1.2	0.9000
Photomultipler Tube, Cesium Antiminide CsSb Photocathode	0.25	0.75	0.5000
Photomultipler Tube, Multialkali (Na-K-Sb-Cs) Photocathode	0.3	0.93	0.6300
Photomultipler Tube, CsI Photocathode	0.2	0.9	0.7000
Photomultipler Tube, S1 Type Photocathode	0.3	1.2	0.9000
Photomultipler Tube, Cesium Galium Arsenide (CsGaAs) Photocathode	0.3	0.93	0.6300
Photomultipler Tube, (CsInGaAs) Photocathode	0.3	1	0.7000
Silicon (Si) CCD	0.2	1.1	0.9000
Indium Gallium Arsenide (InGaAs) CCD	0.68	1.7	1.0200
Visible Range	0.36	0.78	0.4200

of the electrical output (or detector efficiency or sensitivity) as a function of optical energy input. This relationship is expressed as follows:

$$R = Q\left(\frac{\lambda(\mu m)\cdot A}{1.23985(\mu m\cdot W)}\right) = \text{Amps/Watt} \tag{3.33}$$

This computational form is derived from the responsivity expression, equivalent to the quantum efficiency times the electron charge divided by Planck's constant times the frequency of the optical signal; the responsivity changes as a function of the wavelength of the incident energy. Note the terms $\lambda(\mu\text{m})$ and (μm) refer to the wavelength of light/energy in microns; these terms cancel in practice.

Note that responsivity (R) in Amps/Watt may be converted to Quantum Efficiency (Q) on a scale of 0 to 1 and is expressed as (Watts · Wavelength (in μm))/Amps. This conversion of R to Q may be expressed as:

$$Q = \frac{1.23985R\cdot W}{A} \tag{3.34}$$

3.1.2.3 *Theoretical Limit of D* for a Detector System*

Noise from a detector and its amplifier electronics often results from multiple factors, including temperature effects. However there is a theoretical noise limit associated with the background fluctuation attributable only to the detector, irrespective of

whether it is a photovoltaic or photoconductive detector. The limits of D* are somewhat different for photovoltaic or photoconductive detectors and may be expressed in terms of the wavelength of light (λ), Quantum efficiency (Q), and the flux of background radiation (F). The constants for Planck and the speed of light are also included in the expressions.
For the photoconductive detector:

$$D^* = \frac{\lambda\sqrt{Q}}{hc\sqrt{2F}} \text{ in units of cm} \cdot \text{Hz}^{1/2}\text{W}^{-1} \quad (3.35)$$

For the photovoltaic detector:

$$D^* = \frac{\lambda\sqrt{Q}}{2hc\sqrt{F}} \text{ in units of cm} \cdot \text{Hz}^{1/2}\text{W}^{-1} \quad (3.36)$$

This D* value is defined as the detectivity of a radiation detector as a function of the wavelength, the square root of the quantum efficiency (Q), and the background radiation flux (F). The background radiation (F) is computed in Watts per area-squared (e.g., cm^2) based on the detector temperature, typically expressed in degrees Kelvin. In computing the *Signal to noise for a Detector System*, a higher shunt impedance indicates a lower overall dark noise for the detector such that a detector with the same peak response, and responsivity profile, with higher shunt resistance will yield a higher signal to noise for measurement; a higher signal to noise is preferable in most cases.

3.1.2.4 *Gratings*

When incident light strikes a diffraction grating, the light is separated into its component wavelengths with each wavelength scattered at a different angle. Formulas are available to calculate the angles of diffraction, grating efficiency, intensity distribution, and resolution. To calculate the particular angle (θ_d) at each wavelength of light (λ_a) which is scattered from a diffraction grating, the following expression is applied.

$$\theta_d = \sin^{-1}\left(\frac{k \cdot \lambda_a}{s}\right) \quad (3.37)$$

Where θ_d is the angle of diffracted light from a normal (i.e., perpendicular to the plane of the diffraction grating) angle; k is the order number (integers such as 0, 1, 2, 3, ... ,k); s is the spacing between each line on the grating (in the same units as the wavelength (λ_a) used in the equation, e.g., microns, nanometers, etc.); and λ_a is the wavelength of the incident light in air (for white light it represents the wavelength of interest for a calculation of the diffraction angle following dispersion). Note that for zero-order light there is no dispersion; but only straight line reflection of the incident light (i.e., mirror reflection); and thus for zero-order; $\theta_d = \sin^{-1}(0) = 0$.

The dispersion of a grating refers to how broadly the monochromator disperses (or spreads) the light spectrum at the sample specimen position. Dispersion is generally expressed in units of nanometers per millimeter (nm/mm). Dispersion depends upon the groove density (number of lines or grooves per millimeter) on the grating surface. The light collection of the grating depends upon its overall surface area and surface shape (e.g. parabola, aberration corrected, concave, etc.).

The intensity distribution of light (I) from the surface of a diffraction grating is given by the expression:

$$I = \left(\frac{I_0}{S^2}\right)\left(\frac{\sin b}{b}\right)\left(\frac{\sin S \cdot a}{\sin a}\right) \tag{3.38}$$

where I is the intensity from the surface of the grating; I_0 is the incident energy intensity; S is the number of slits the light must pass through following the diffraction; b is designated as $(\pi \cdot s/\lambda_a)$, where s is the spacing of the lines on the grating (to simplify the calculation use the same units as the wavelength (λ_a) used in the equation); λ_a is the wavelength of the incident light for a calculation of the diffraction angle; and a is designated as $(\pi \cdot \sigma/\lambda_a)$, where σ is the slit width in the same units as λ_a.

The resolution (P) or a diffraction grating is given by the expression as follows:

$$P = k \cdot n = \frac{\lambda}{\Delta\lambda} \tag{3.39}$$

where k is the order of the interference pattern; n is the total number of lines on the diffraction grating surface; λ is the wavelength at which the resolution is determined; and $\Delta\lambda$ is the distance in wavelength space between the lines or optical phenomenon to be resolved.

The fundamental grating equation is given by:

$$sin\,\alpha + sin\,\beta = 10^{-6} k \cdot n \cdot \lambda \tag{3.40}$$

where n is the grating groove density per mm. In most monochromators, the location of the entrance and exit slits are fixed and the grating rotates around a plane through the center of the face. The angle, θ_V, is a constant determined by:

$$\theta_V = \beta - \alpha \tag{3.41}$$

If the value of α and β is to be determined for a given wavelength (λ), the grating equation may be expressed as:

$$kn\lambda \cdot 10^{-6} = 2\sin\left[\frac{\beta+\alpha}{2}\right]\cos\left[\frac{\beta-\alpha}{2}\right] \tag{3.42}$$

To calculate α in a Monochromator use:

$$\alpha = \sin^{-1}\left[\frac{10^{-6}kn\lambda}{2\cos\left(\frac{\theta_V}{2}\right)}\right] - \frac{\theta_V}{2} \tag{3.43}$$

The symbols for the grating equation are as follows. Please note that an illustration of multi-order diffraction is shown in Figures 3.7, 3.8a, and 3.8b. For the symbols in the grating equation note that α is the angle of incidence (in degrees); β is the angle of diffraction (in degrees); k is the diffraction order number (i.e., integers of 0, 1, 2, etc.); n is the groove density (in grooves per mm); θ_V is the angle of deviation or dispersion (in degrees); and λ is the wavelength of light used to compute the dispersion of interest (in nm).

3.1.2.5 *Beam Splitters*

Beam splitters are optical devices used to divide and recombine an optical beam (or beams) of light. They can be produced using half-silvered (half-aluminized)

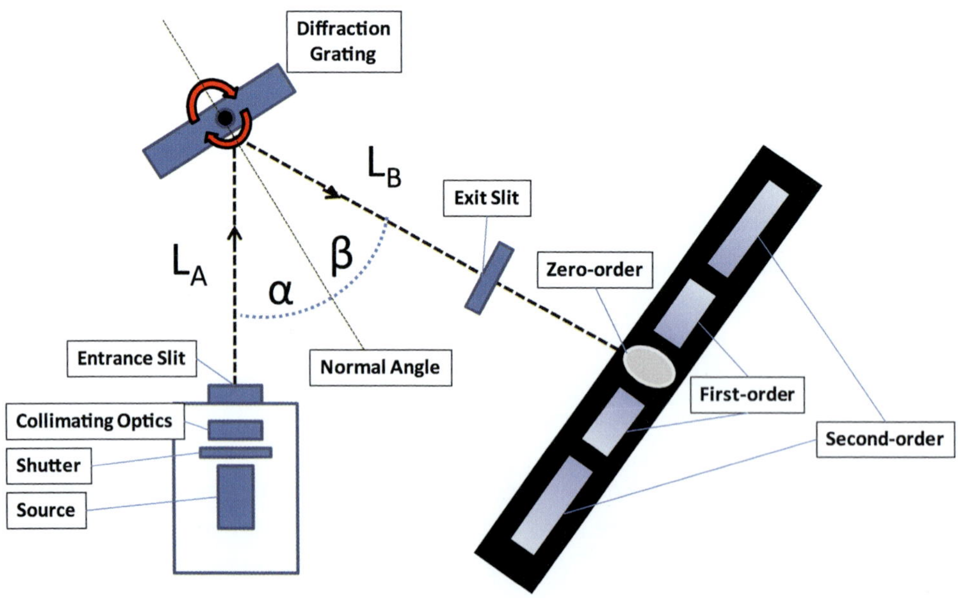

Fig. 3.7 Schematic showing multiple order dispersion from a pre-dispersive monochromator grating system. Note that angle α is equal to angle β.

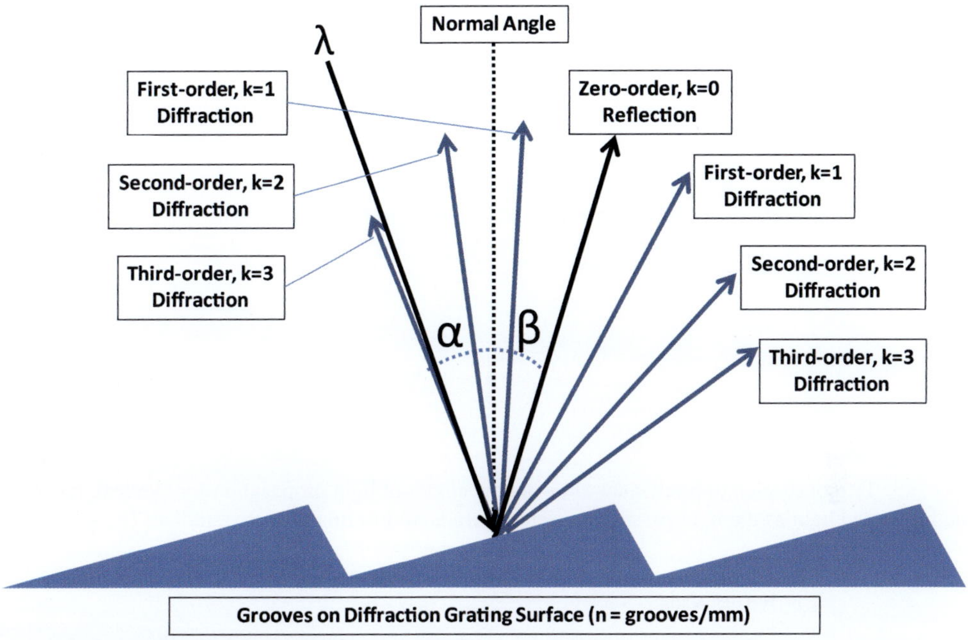

Fig. 3.8a Schematic showing grating surface and the resulting multiple order dispersion from the grating optical surface following the grating equation. Note here that angle α is equal to angle β.

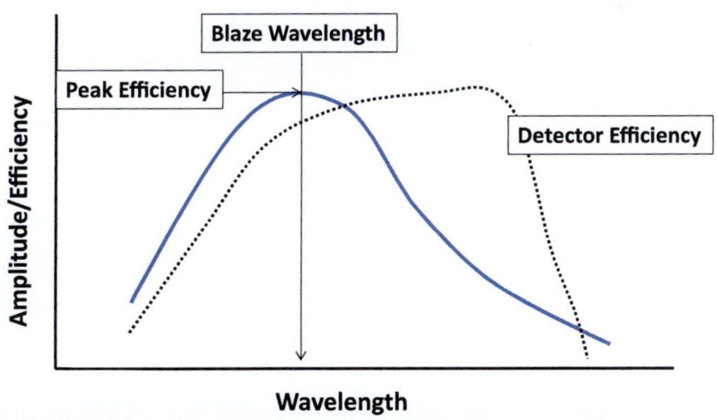

Fig. 3.8b Illustration showing the relationship between a typical diffraction grating efficiency curve (solid line) and photo detector efficiency as photomulitiplier tube (dotted line). Also shown is the concept of the peak efficiency of the grating at the grating blaze wavelength.

mirrors that reflect approximately 50% of the energy incident to them; the remaining 50% is transmitted through the beam splitter; thus the term "beam splitter" is used. This is what a half-mirror refers to as approximately half the light is transmitted and half of the light is reflected.

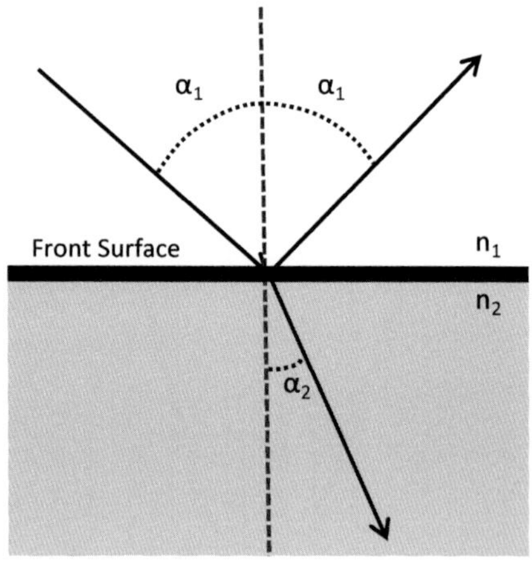

Fig. 3.9 Demonstration of Snell's law showing the angles of light propagation for incident, reflected and refracted light as the light passes between materials of differing refractive indices (n_1 and n_2).

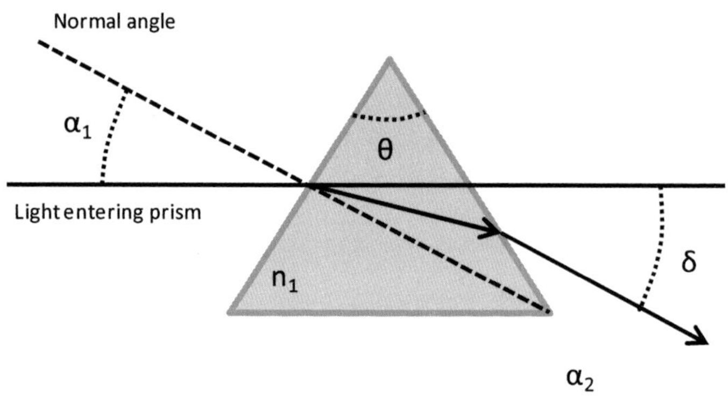

Fig. 3.10 Angles of wave propagation through an equilateral prism showing dispersion of light.

3.1.2.6 *Prisms*

Prisms may be used to disperse light into its spectral components, or as right-angle prisms with reflective coatings on the hypotenuse side to bend light at a 90° angle (i.e., mirror, reflective, diagonal prism). The dispersive properties of prisms have been known since the late 17th century. The deflection or dispersion of a prism is described using Snell's law at each optical surface of the prism, considering the refractive index of the prism at each wavelength. Snell's law describes the change in the path of light crossing an interface between two different materials (with different refractive indices) when the incident angle of the first surface is other than 90°. The wavefronts or wave propagation angle through the interface must move toward the normal angle as shown in Figure 3.9, with the numerical explanation of Snell's law given by:

$$n_1 \sin \alpha_1 = n_2 \sin \alpha_2 \tag{3.44}$$

where n_1 is the refractive index of the first medium at the interface of two materials of varying refractive indices; n_2 is the refractive index of the second medium at the interface of two materials of varying refractive indices; α_1 is the angle of incidence and reflection of the impinging light onto the surface of the second medium; and α_2 is the angle of refraction of the light passing through the interface of two materials of varying refractive indices.

For an equilateral (dispersive) prism, the wave propagation angle (δ) (shown in Figure 3.10) through the prism is given by:

$$\delta = \alpha_1 + \sin^{-1}[(\sin\theta)(n^2 - \sin^2\alpha_1)^{1/2} - (\sin\alpha_1 \cos\theta)] - \theta \tag{3.45}$$

where α_1 is the light angle incident to the surface of the prism; θ is the wedge angle of the prism (60° for an equilateral triangle prism); n is the refractive index of the prism at the wavelength of incident energy; and δ is the wave propagation angle (angle of refraction) through the prism.

3.1.2.7 *Polarizers*

A diversity of polarizing elements have been developed for the purpose of rotating or selecting light of a specific electronic vector orientation. When the electronic direction vector of light incident to a surface is parallel to the electronic field vector of that surface, increased interaction of the incident light (as energy absorption) occurs. The principle is important in characterizing the surface chemistry for optical components, thin films, metal surfaces, and semiconductor interfaces. The use of a quartz plate can act as a polarizer that will rotate the plane of linearly polarized light. This rotation (P) can be described using the following relationship:

$$P = \pi \cdot t(|n_\Lambda - n_P|)\lambda^{-1} \tag{3.46}$$

where t is the thickness of a quartz plate cut perpendicularly to the optical axis (in nm); n_Λ is the refractive index for right circularly polarized light; n_P is refractive index for left circularly polarized light; and λ is the vacuum wavelength of light entering the quartz plate (in nm). Note: for quartz $n_\Lambda = 1.5582$ and $n_P = 1.5581$. Please note the index section on polarization measurements for more details.

3.1.2.8 *Electronic Components used in Spectrometry*

For references describing the electronic components of optical spectrophotometers, see modern textbooks on electronic instrumentation including references 4–6. Other volumes in this set discuss more details regarding electronic components for instrumentation, please see index for specific topics. Two special categories of electronic devices deserve mention when discussing spectrometers. The first category includes detectors and detector electronics that produce a current or voltage signal proportional to the amount of energy striking the detector. Detector stability is provided by appropriate electronic circuitry allowing the detector signal to be selectively amplified with the minimum distortion or background noise, thus electronic circuitry enhances the signal-to-noise ratio of the detector signal.

Digital microcomputers comprise the second essential electronic element for modern spectrophotometers. With the addition of appropriate software, sophisticated instrument control, and data processing, the usefulness and user friendliness of instruments can be made faster, and more accurate. A basic overview of electronic components is shown in Figure 3.11, and their associations with instrumentation as shown in Figures 3.1 through 3.5 within this chapter.

3.1.3 *Properties of Spectrophotometers*

3.1.3.1 *Aperture Diameter*

To calculate the aperture (a) required by an optical system to resolve two objects with known linear separation, the Raleigh criterion for resolution is used as:

$$a = 1.22\lambda \cdot \alpha_R^{-1} \tag{3.47}$$

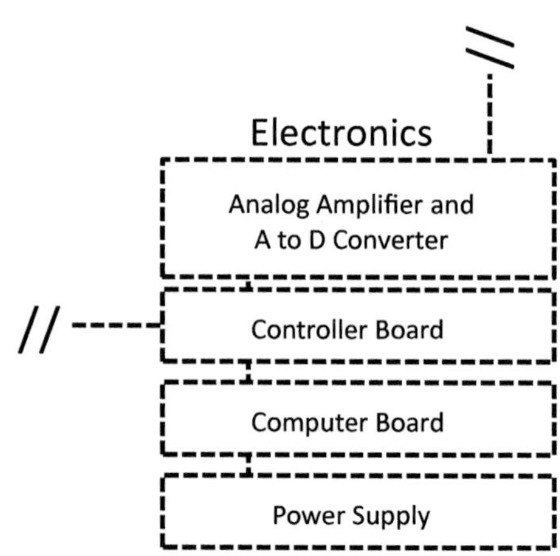

Fig. 3.11 Basic electronic systems for spectrometer systems.

where α_R is the angle of separation from the measuring device exit aperture to the objects to be resolved. (This is calculated as tan α = opposite/adjacent, where α_R is expressed in radians); λ is the wavelength of light observed from the objects; and a is the aperture of the optical system.

3.1.3.2 *Entrance and Exit Pupils*

The entrance pupil refers to the size and location of the entrance aperture between the light source and the remainder of an optical system. The exit pupil refers to the size and location of the exit aperture within an optical system or optical train just prior to the detector. Both the entrance and exit pupils directly affect the energy throughput and spectral resolution of a spectrophotometer.

3.1.3.3 *Band Pass (Bandpass) and Resolution*

The terms bandpass and resolution are used to express the capability of a spectrometer to separate spectral bands or lines which are separated by some finite distance on the spectral energy scale (i.e., frequency or wavelength scale). For an instrument that disperses energy over a prespecified spectral region of the electromagnetic spectrum the bandpass of a spectrometer is used to describe which portion of the spectrum can actually be isolated by the spectrometer in a "pure" wavelength form. The spectrometer bandpass is dependent upon the dispersion of the grating and the entrance and exit slit widths.

An illustration is often used to demonstrate the problem associated with measuring monochromatic light using conventional spectrophotometers. If the ideal spectrometer was used to measure a bright line emission spectrum at a single wavelength (λ_1), the spectrum would appear as a single line (Figure 3.12).

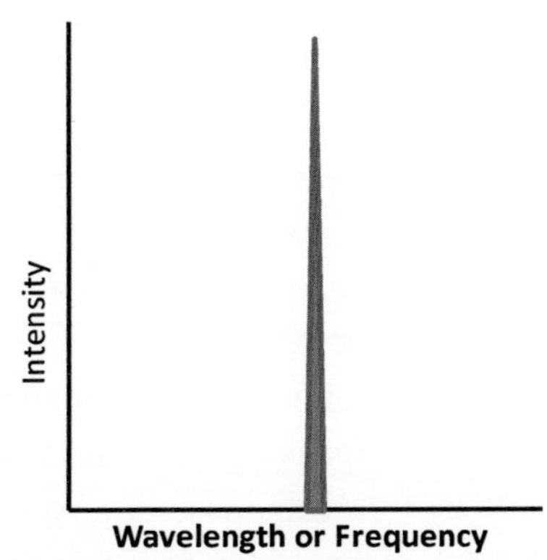

Fig. 3.12 The bright line emission spectrum at a single wavelength as it would appear in an ideal spectrophotometer.

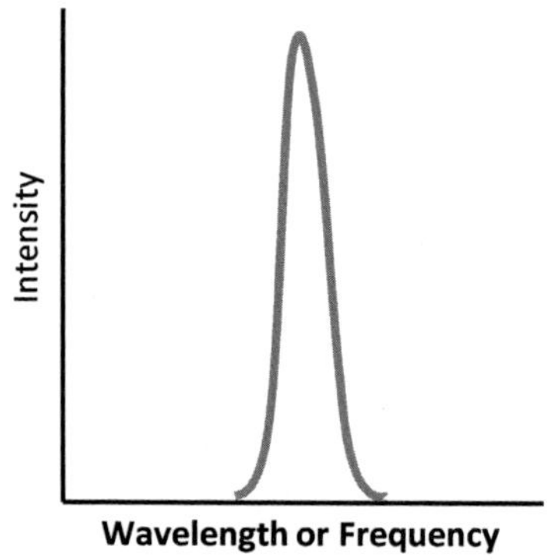

Fig. 3.13 Spectrum of a bright line emission source (e.g. deuterium lamp). The characteristic broadening is an illustration of the bandpass or resolution of a spectrophotometer.

What really occurs when such a spectrum is measured using a conventional spectrometer is a broad band spectrum as shown in Figure 3.13. The spectrum assumes a Gaussian-like (or bell-shaped) curve, actually closer to Lorentzian. This characteristic broadening of a line spectrum through the spectrometer is an illustration of the spectrometer bandpass. The actual bandpass for any instruments assigned a value by determining the Full Width at Half Maximum (FWHM) height of the bell-shaped spectrum. Thus for the band in Figure 3.13, the FWHM could be empirically determined by finding the wavelength where maximum intensity occurs, and measuring the peak height at this position. This height measurement is divided in half and the bandwidth measured at this height on the band as illustrated in Figure 3.14.

The actual shape of a band is the result of several instrumental characteristics including the overall quality of the optics and detector systems, as well as the width and positions of the entrance and exit slits. Every dispersive spectrometer consists of a dispersive element (e.g., diffraction grating or prism) in combination with an entrance and an exit slit. The image of the entrance slit and exit slit determines the spectrometer bandpass which is sometimes referred to as the slit function. Actually the slit function is the result of the combination of the images of these two slits. The bandshape of a dispersive spectrometer is shown in Figure 3.14. Other factors associated with optical and electronic quality cause a rounded overall shape. The bandpass of a spectrometer is equal to the FWHM. Often texts dealing with instrumentation will site that the bandpass of a spectrometer is approximated by the product of the linear dispersion of the monochromator and the entrance or exit slit width (whichever is larger).

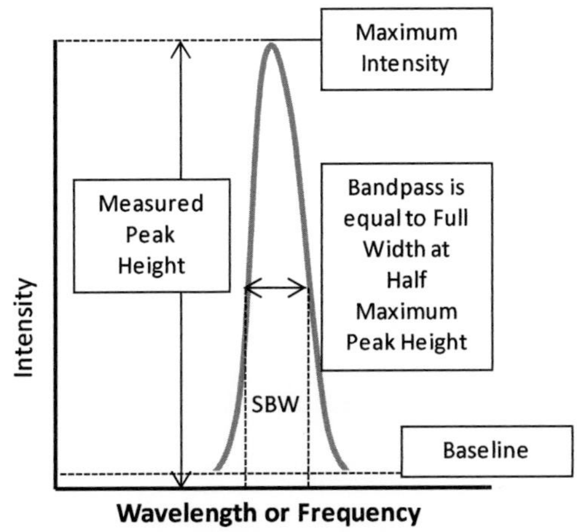

Fig. 3.14 Illustration of the determination of Spectral Bandwidth (SBW), also known as Resolution Calculation, Full Width at Half Maximum (FWHM), Spectral Bandpass, Spectral Linewidth, and Bandwidth. Using the bell-shaped peak obtained by using a bright line source projected through a monochromator optical system, or another isolated absorbance band, the width at half the measured height is computed as the Resolution Calculation equal to the SBW.

The resolution of a spectrometer can be defined as the minimum distance between two peaks that can be detected by the spectrometer under designated operational performance settings. Resolution is calculated by multiplying the slit width (generally expressed in mm) by the dispersion of the monochromator (in nm per mm). Due to practical issues and non ideal optics, the actual resolution of a spectrometer must be slightly less (poorer) than the theoretical value.

To summarize, bandpass and resolution are identical in practice. Only the resolution specification of a spectrometer is the expression of bandpass under the specified measuring conditions of an instrument dependent upon the slit width settings.

The empirical resolution of a spectrometer is manually determined by measuring the FWHM in mm for two narrow bands which are completely resolved (to the baseline) using the spectrometer and projected onto a graphical format. The spatial difference between the maximum absorbance (λ_{max}) is determined between the bands (in mm); simultaneously noting the difference between the λ_{max} points in nm. The various measurements required for this calculation are shown in Figure 3.15 and illustrated by the relationship:

$$\text{The bandpass} = \text{resolution} = \frac{\text{band.difference.in.nm}}{\text{band.difference.in.mm}} \times \text{FWHM in mm} \quad (3.48)$$

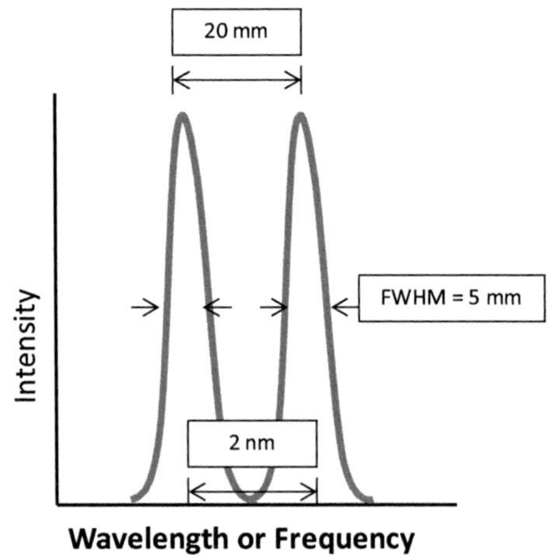

Fig. 3.15 The manual method used to determine maximum resolution of a spectrometer under specific measurement conditions. In the example shown Bandpass = Resolution = FWHM × Dispersion = 2 nm/20 mm × 5 mm = 0.50 nm resolution. The mm measurements refer to a graphical chart made of the spectral lines.

3.1.3.4 *Numerical Aperture*

The numerical aperture, designated as NA is a measure of how much light can be collected by any optical system. The NA is expressed as the product of the refractive index of the material where incident energy is directed into (n_i) at the wavelength of interest, times the sine of the ray angle maximum difference from normal incidence (note that normal is perpendicular to the optical surface); this function is given as:

$$NA = n_i \cdot \sin \theta_{max} \tag{3.49}$$

3.1.3.5 *Attenuation*

Attentuation is a term often used to describe the light transmittance losses within (or through) an optical system. Losses in transmitted light through spectrometers are due to absorption, reflection, scattering, and optical misalignment; the losses can vary with temperature and wavelength. The quantity of optical loss can be expressed as an attenuation rate in decibels (dB) of optical power per unit distance (cm). The usual attenuation occurs through inefficiencies in source launch optics, lenses, mirrors, bandpass filters, slits and J-Stop features, detector focus, fiber optic losses (when applicable), and the like. Typical changes in the energy transmitted through a spectrometer over time result from source aging, launch optics alignment, temperature variations resulting in mechanical alignment changes, inefficient optical couplings within an optical path, aging of mirrored surfaces and lens optical coatings, and typical soiled optical surfaces (e.g., lenses, mirrors, and windows).

The losses in energy transmitted through a spectrometer can be calculated by using Beer's law, which states that the irradiance of the energy through an absorbing medium falls exponentially with the distance of transmission through a medium. The relationship for this is as:

$$\mathrm{I_d} = \mathrm{I_0} \cdot 10^{-\alpha \mathrm{d}}, \text{ and it follows that} \tag{3.50}$$

$$\mathrm{T} = \frac{I_d}{I_0} = 10^{-\alpha d}, \quad \text{and} \tag{3.51}$$

$$-\log\left(\frac{I_d}{I_0}\right) = \alpha d, \quad \text{and so} \tag{3.52}$$

$$\alpha = -\log\left(\frac{I_d}{I_0}\right) \cdot d^{-1} \tag{3.53}$$

where $\mathrm{I_d}$ is the irradiance measured at distance (d) from the source through the optical system (train) in cm; $\mathrm{I_0}$ is the source irradiance at $\mathrm{d} = 0$ from the source; and α is the attenuation loss (absorption) coefficient in units of dB/cm through the optical system. Attenuation losses are wavelength dependent; thus the value for α is a function of the incident wavelength and its interaction with the optical system.

3.1.3.6 *Etendue*

The etendue (or relative throughput advantage) for an optical system is the product of the potential illuminated surface area (A) times the solid angle of the optical system. Traditionally, this is represented by the following equations, where ε' represents the etendue and Ω_S represents the solid angle. Thus the etendue formula is given as:

$$\varepsilon' = \mathrm{A} \cdot \Omega_S \tag{3.54}$$

and the solid angle is given by the relationship:

$$\Omega_S = 2\pi \left(1 - \sqrt{(1 - (NA)^2)}\right) \tag{3.55}$$

Therefore, the previous equations allow us to calculate the relative improvement for an optical spectrometer. As can be seen, the NA and aperture diameter are preeminent factors for designated throughput for any optical system as shown in Table 3.6.

3.1.3.7 *Relative Throughput*

The relative throughput (T) of an optical system represents the overall effectiveness of an optical system to transmit light relative to the amount of energy introduced

Table 3.6 Etendue and relative throughput as a function of Numerical Aperture (NA).

NA (Numerical Aperture)	Relative etendue (ε') for a 1-mm-diameter aperture	Relative throughput
0.20	0.10	1
0.40	0.40	4
0.60	1.00	10

into the system (I_0) from the optical light source of the spectrophotometer. It is defined as the ratio of light energy passing into an optical system to the light energy passing out of the optical system. For dispersive spectrometers, this relationship is defined as:

$$T = \frac{\pi D w_s \Sigma_g (R_1 \cdot R_2 \cdot \ldots \cdot R_k)}{4f^2} \tag{3.56}$$

where D is the dispersion constant of the diffraction grating (in nm/mm); w_s is the exit slit width in mm; f is the f/number of the optical system and is designated by (f/number = 1/2(NA)); R_k is the reflectivity of mirrors or the transmittance of lenses at a particular wavelength or set of wavelengths, in terms of 0.0 to 1.0 where 1.0 represents 100% reflectance or transmittance; and Σ_g is the spectral efficiency of the grating (i.e., approximately 0.80 at the blaze wavelength, but this will vary with grating design and can be either calculated or measured.)

3.1.3.8 *Signal-to-Noise Ratio*

The theoretical total signal (S) from an optical system can be given by the product of a number of optical parameters as:

$$S = R_S B_\lambda \varepsilon' \tau Q \tag{3.57}$$

where R_S is the light source spectral radiance; B_λ is the spectral bandwidth; ε' is the etendue of the spectrometer optical system; τ is the combined transmittance and emissivity losses; and Q is the quantum efficiency.

The measured signal-to-noise (S/N) from an optical system indicates the maximum amount of signal produced by an instrument as compared to the noise of the instrument during measurement. The higher the number, the better the quality of the instrument. The s/n ratio can be calculated from the full transmitted (or reflected) signal divided by the RMS (root means square) noise (in Transmittance or Reflectance units). Thus for a 100% line with RMS noise as 0.001 percent, the s/n = 100/0.01 = 10,000:1. This applies when RMS noise for a transmission measurement is calculated empirically as:

$$\text{RMS} = \sqrt{\frac{1}{n}\sum_{i=1}^{n}(T_i - \bar{T})^2} \quad (3.58)$$

where T_i = the individual Transmission value at each data channel i; and $\bar{T}$ is the mean (or true) transmission value for each data channel. The corresponding s/n for reflectance measurements is calculated by substituting reflectance for transmission.

3.1.3.9 *Dynamic Range and Stray Radiant Energy*

The range of a specified component concentration over which a sensor response is directly proportional to a change in concentration is a definition for the dynamic range of a spectrometer. Dynamic range is limited by both stray light and noise. A basic understanding of Beer's law allows one to calculate the maximum theoretical dynamic range for an instrument and the relationship between stray light and the maximum observable absorbance value for a spectrometer. The following relationships allows the error in a photometric measurement due to stray light to be computed.

Beer's law is given as the true absorbance measurement is equal to:

$$A_\tau = \log_{10}\left(\frac{I_0}{I_t}\right) \quad (3.59)$$

where A_τ is the true absorbance value for a sample; I_0 is the incident intensity measured for a true spectrometer; and I_t is the true intensity of a sample measured with the spectrometer. To follow this rationale, we compare a measured absorbance value to the true absorbance value considering stray light. In this case, stray light is defined as any light impinging upon the detector that is not monochromatic light. In this case, the effect of stray light on the true absorbance is given by:

$$A_m = \log_{10}\left(\frac{I_0 + I_s}{I_t + I_s}\right) \quad (3.60)$$

where A_m is the measured absorbance value for a sample; I_0 is the incident intensity measured for the spectrometer; I_t is the true intensity of a sample measured with a true spectrometer; and I_s is the stray light intensity (as a fraction of the incident energy as I_0 = 1.0). The effect of stray light on a measurement is summarized in Volume 2, Chapter 3, Sections 3.1.3.8 and 3.1.3.9.

3.1.3.10 *Wood's Anomalies*

R.W. Wood discovered the phenomenon whereby diffraction gratings demonstrate that based on the grating properties (i.e., groove separation of lines on a grating, and the depth of those grooves), there are narrow regions of dispersed energy that change rapidly in the amount of energy diffracted.[7] Lord Rayleigh also observed these anomalies and referred to them as "passing-off orders" of energy.[8]

Light diffracted along the plane of the grating is reflected or dispersed back into the allowed orders of light causing a dramatic increase in energy. The redistributed light is termed "passed off". The additional energy adds to the spectral response with a sharp increase at the critical wavelength (λ_{Woods}) and an equally sharp decrease to the lower wavelength side of the dispersion. The grating efficiency is actually increased at the point of the anomaly as two orders of the dispersed energy are combined. This increased energy is perpendicularly (S-) polarized to the grating lines (see the main index for sections describing details of polarized light).

The wavelength position (in nanometers) of the Wood's anomalies for a diffraction grating is given as Equation (3.61). The $\pm$ represents the order on either side of the grating.

$$\lambda_{\text{Wood's}} = \frac{\delta \cdot (\sin \alpha \pm 1)}{k} \tag{3.61}$$

Where δ is the groove separation (in nanometers); α is the angle of incidence of the light onto the grating (as compared to a normal angle to the grating surface); and k is the scattered order, or the diffraction order of the light dispersed (or diffracted) from the grating surface (as ± 1, ± 2, ± 3, etc,). For example, if $\alpha = 45°$, $\delta = 300$ nm, and $k = 2$ and we wanted to know where the increased energy will occur. The answer is that at 256.05 nm a large energy spike is expected to occur from this specific grating and dispersion pattern, as shown in Equation 3.62.

$$\lambda_{\text{Wood's}} = \frac{300\,nm \cdot (\sin 45° + 1)}{2} = 256.05\,\text{nm} \tag{3.62}$$

3.1.4 *Instrument Comparison and Evaluation Methods*

A number of standard measurements may be completed that allow detailed comparison of overall instrument performance. These are described in detail within other volumes of this work; the reader is referred to the master index.

A series of well-defined tests are used to qualify instrument performance and to determine which issues are problematic due to deficiencies in instrument design features, or in sample presentation systems. These tests are related to alikeness in measurement performance between instruments and to overall accuracy, and precision (as reproducibility and repeatability, respectively). Note that the terms optical density (O.D.) and absorbance units (AU or Au or A) are synonyms. The techniques used for evaluation of spectrometer performance are described in detail within Chapter 6 of this volume, entitled *Data Analysis and Presentation*. Please also refer to other volumes for additional instrument test descriptions.

References

1. ASTM E131-10, Standard Terminology Relating to Molecular Spectroscopy, 2010. ASTM International, 100 Barr Harbor Drive, PO Box C700, West Conshohocken, PA, 19428-2959 USA.
2. P.R. Griffiths and J.A. De Haseth, *Fourier Transform Infrared Spectrometry*, 2nd edn. (Wiley-Interscience, 2007).
3. J.M. Chalmers and P.R. Griffiths (Eds.), *Handbook of Vibrational Spectroscopy, 5 Volume Set*, 1st edn. (Wiley, 2002)
4. D.M. Buchla and W. McLachlan, *Applied Electronic Instrumentation and Measurement*, 1st edn. (Prentice Hall, 1991).
5. J. Turner and M. Hill, *Instrumentation for Engineers and Scientists, Textbooks in Electrical & Electronic Engineering*, Book 8 (Oxford University Press, 1999).
6. G.S. Sawhney, *Biomedical Electronics and Instrumentation Made Easy*, 1st edn. (I. K. International Publishing House, 2011).
7. R.W. Wood, On a remarkable case of uneven distribution of light in a diffraction grating spectrum, *Philos. Mag.* **4** (1902) 396–402.
8. Lord Rayleigh, Note on the remarkable case of diffraction spectra described by Prof. Wood, *Philos. Mag.* **14** (1907) 60–65.

Chapter 4

Sampling Considerations for Ultraviolet Spectroscopy

Contents

4.1 Introduction

Like other spectroscopic techniques, the quality of an ultraviolet spectrum depends upon the consistency in sample preparation and the reproducibility of the optical interface between the sample and the spectrophotometer. A variety of sampling accessories are commercially available to optimize measurements and improve repeatability and reproducibility of spectral data. The typical ultraviolet sampling techniques generally involve transmission measurement of liquids, but as this volume describes there are many other sampling techniques common to ultraviolet spectroscopy, including fluorescence and phosphorescence.

This chapter is intended to serve as a guide to provide basic information on sampling methods for ultraviolet and fluorescence spectroscopy. The best advice for sample preparation is to use a simple, clean method with careful documentation of each preparation step. For ultraviolet measurements a simple diffuse or specular reflection measurement will provide the best quality data for opaque samples, while a transmittance method is optimum for clear or transparent solids or liquids. The actual choice in measurement geometry may well depend on the final application referring to qualitative or quantitative analysis. Samples used for quantitative or qualitative analysis must be prepared as other samples used for collection of an original and reproducible database. The database, or spectral library, is used for quantitative calibration modeling or for qualitative comparisons. These algorithms and methods are described in more detail within Chapter 6 of each volume within this five volume set.

As with other spectroscopic regions, the use of spectral data pre-processing should be applied with intentional purpose. Pre-processing steps, such as baseline correction, smoothing, derivatives, normalization, or scatter correction are useful when applied appropriately. It is important to appreciate the impact and the consequences of the specific data manipulation used. Distortions may be introduced into an ultraviolet spectrum that can lead to errors in quantitative measurements or spectral interpretation. It is essential to keep detailed records of the specific form of pre-processing used. Refer to Chapter 5 and the index for recommendations on instructions for spectral acquisition, data and presentation formats, and comments regarding spectral data manipulations. Chapter 6 of each volume describes mathematical procedures used to process spectroscopic data.

4.1.1 *General Considerations*

Ultraviolet energy is a form of electromagnetic radiation nominally extending from the vacuum ultraviolet (approximately 190 nm) to just before the visible region — practically defined as approximately 360 nm (i.e., 0.36 microns). In this chapter, this spectral region will be described as ultraviolet or UV radiation. Most ultraviolet interactions with matter are recorded as absorptions of energy, either as direct or indirect measurements and use Beer's law relationship to convert transmittance or reflectance measurements to the absorbance scale.

When an excited molecule re-emits absorbed ultraviolet energy (usually as lower energy visible light) the molecule is said to fluoresce. Fluorescence occurs as rapid emission in 0.5×10^{-9} to 20×10^{-9} sec. Phosphorescence occurs when absorption of ultraviolet energy requires minutes or hours to fully re-emit absorbed energy and return to ground state. See Section 2.1.4 in this volume for more details.

As with other spectroscopic methods, ultraviolet measurements may be made via light transmittance through samples, or by reflection directly from a ground (or powdered) sample surface (Figures 4.1 and 4.2). The absorption properties of molecules using ultraviolet illumination is much less than in the visible or infrared spectral regions. In other words, the absorption coefficients for most molecules is

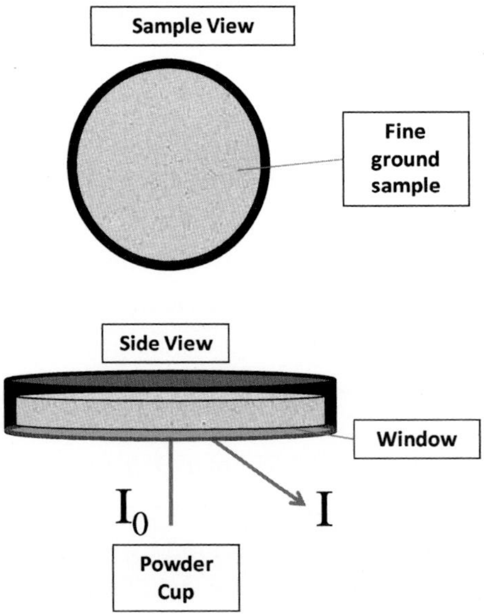

Fig. 4.1 UV diffuse reflectance sample cup configuration. In diffuse reflection, there is an interaction of the photons with the molecules of the sample, such that absorption occurs. This absorption is converted from reflectance to absorbance for computing an absorbance spectrum.

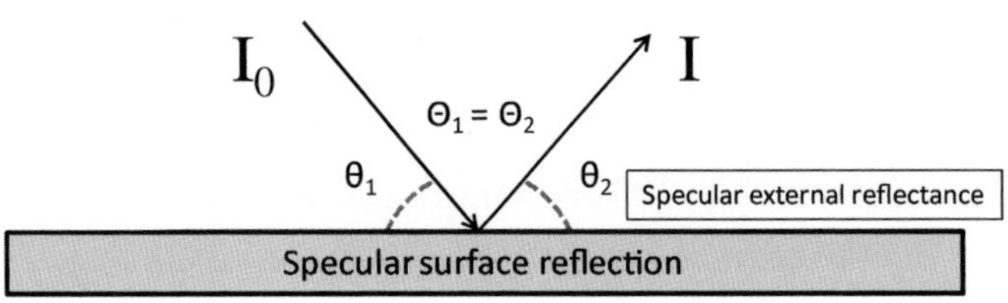

Fig. 4.2 UV specular reflectance sampling configuration. Angle of illumination and reflection are equivalent as shown by θ_1 and θ_2. This reflection is elastic and contains little or no real chemical information from the sample. For elastic scattering/reflection the wavelength (and energy) of the incident photon is maintained (or conserved) and only the photon direction is changed.

much smaller in the ultraviolet region than for other molecular spectroscopy spectral regions.

It is assumed that the reader has a basic knowledge of optical absorption and emission spectroscopy and the molecular theory of chemistry. Ultraviolet spectroscopy is one of the analytical techniques available for the measurement of absorption of radiation as a consequence of electronic transitions within a molecule. For an understanding of underlying theory, the reader is directed to Chapters 1 and 2 as well as the index of this volume and the main index. Fluorescence is an emission spectroscopy technique involving the use of ultraviolet light for excitation of molecules and resultant fluorescence emission at a lower energy or longer wavelength than the excitation energy. Fluorescence is specifically addressed within this volume in Chapter 2, Section 2.1.4.

The term ultraviolet spectroscopy can be generally applied to measurements from 10 nm to 400 nm. Most commonly, for general laboratory and non-vacuum use, the ultraviolet wavelength region is designated as 190 nm to 360 nm. Note that the ultraviolet region is defined differently depending on the organization using the definitions and the specific application of the ultraviolet region (Table 4.1). For example, the ultraviolet region has been defined as from 285 nm to 400 nm, with the visible region defined as 400 nm to 750 nm. The ultraviolet region is then further defined as being either UV-A with radiation of wavelengths from 315 nm to 400 nm, or UV-B with radiation from 285 nm to 315 nm.[1] A standard document from ASTM International defines the ultraviolet region from 200 nm to 400 nm.[2] Other organizations or methods define the UV region as from 185 nm[3] to above 380 nm. See Volume 2, Chapter 7, Section 7.2.1 for details on the visible spectral region, including color analysis. Further detailed definitions of the ultraviolet region are given in ISO-21348 (Table 4.1).[4–5]

The lower wavelength limit is normally defined by the sampling method or matrix or by the optics of the instrument. In the latter case, the standard detector for the ultraviolet region is a photomultiplier tube (PMT) for highest signal to noise and fastest response. A more comprehensive list of detectors most useful for ultraviolet measurements include photomultiplier tubes (PMTs, 160–1100 nm); charge-coupled devices (CCDs, 180–1100 nm); and silicon-based photodiode arrays (180–1100 nm). The lower limit of detection at near 190 nm is due to the absorption of the ultraviolet energy from the electronic transitions of the atomic and molecular oxygen (O_2) and nitrogen (N_2) from ambient air. If a vacuum is used for the sample

Table 4.1 ISO-21348 definitions for the ultraviolet region.

UV region name	Abbreviation	Wavelength range (nm)
Ultraviolet	UV	100 nm–400 nm
Vacuum Ultraviolet	VUV	10 nm–200 nm
Extreme Ultraviolet	EUV	10 nm–121 nm
Hydrogen Lyman-α	H Lyman-α	121 nm–122 nm
Far Ultraviolet	FUV	122 nm–200 nm
Ultraviolet C	UVC	100 nm–280 nm
Middle Ultraviolet	MUV	200 nm–300 nm
Ultraviolet B	UVB	280 nm–315 nm
Near Ultraviolet	NUV	300 nm–400 nm
Ultraviolet A	UVA	315 nm–400 nm

and measurement areas, lower wavelength ultraviolet measurements may be made to the low wavelength end of the detector to as low as 10 nm.

The ultraviolet region is used to measure electronic transitions in molecules and their associated atoms. The types of electronic transitions excited using ultraviolet light include nonbonding electrons, electrons in single bonds, and electrons involved in double or triple bonds. These may be excited to several excited energy levels (or states). The ultraviolet spectral distinction between different molecules is found in the principle that attached moieties affect the ability of electrons to "jump" to higher energy states. This ability is affected by the presence of double bonds, conjugations, and the presence of atoms such as oxygen, bromine, and others having pairs of nonbonding electrons. As a consequence, most elements with ultraviolet absorptions have specific wavelengths at which their peak absorbances occur. These peaks may be used to identify a particular molecule or chemical type.

The ultraviolet region is not as "rich" in chemical information related to a molecule as is the infrared region. However, UV spectroscopy reveals enough detail to enable a comparison of materials with a previously identified substance (qualitative analysis or identification). A common use of this property of ultraviolet spectra is found in the pharmaceutical industry, where ultraviolet detectors are frequently employed with High-Performance Liquid Chromatography (HPLC) instruments as a final check before a drug product is released for sale to consumers. The various moieties or chromophores associated with ultraviolet absorption include Nitriles (R-C≡N), 160 nm; Acetylenes (-C≡C-), 170 nm; Alkenes (>C=C<), 175 nm; Alcohols (R-OH), 180 nm and 175-200 nm; Ethers (R-O-R'), 180 nm; Ketones (R-C=O -R'), 180 nm and 280 nm; Amines, primary ($R\text{-}NH_2$), 190 nm and 200-220 nm; Aldehydes (R-C=O–H), 190 nm and 290 nm; Carboxylic acids (R-C=O–OH), 205 nm; Esters (R-C=O–OR'), 205 nm; Amides, primary (R-C=O –NH_2), 210 nm; Thiols (R-SH), 210 nm; Nitrites ($R\text{-}NO_2$), 271 nm; and the Azo-group (R-N=N-R'), 340 nm.

In many cases, factors such as specifications, manufacturer, price, convenience of operation, optical performance/efficiency, and final application requirements and features may be the final arbiter in the selection of a sampling method. The actual application can be an important deciding factor. Sampling techniques may be manual or automated. For additional detailed study, one is referred to references.[6–11]

4.1.2 *Sample Type Versus Sampling Method*

Ultraviolet spectroscopy is most often used to measure clear or transparent solids or liquids; or in some cases opaque solid samples, measured in transmittance and diffuse or specular reflectance, respectively. Fluorescence may be used on liquids or solid samples and is easily adapted to both sample phases. As will be mentioned throughout this text, it is essential that the analyst carefully document all experimental procedures, sampling methods and conditions, and conscientiously note any abnormalities discovered during sample preparation. These details should be recorded in appropriate lab analysis notebooks, either hard copy or electronic.

4.1.2.1 *Liquids*

Ultraviolet spectroscopy is compatible with low-OH (i.e., low water) and with low gas entrapped SiO_2 as processed natural quartz, glass, and silicate cell windows. The presence of OH or entrapped air will limit transmittance below approximately 235 to 240 nm. Table 4.2 lists the most ideal materials for sample cells used in laboratory ultraviolet measurements.

Traditionally, transparent liquids have been considered to be relatively easy to measure using UV spectroscopy. The exception to this is for the measurements of volatile, low-molecular-weight compounds. Such compounds must be measured in sealed vials and rapidly handled so as not to evaporate. The main method for storing liquids during measurement using ultraviolet spectroscopy is an open or sealed transmittance cell or cuvet/cuvette. For quantitative measurements, liquid cells should be measured at a relatively constant temperature of ±0.1°C with a pathlength variation of no more than 2% relative. Alternative to the application of a constant or fixed pathlength cell is a cell whose pathlength has been carefully calibrated using either manual dimensional measurements (caliper), or a variety of absorbance techniques where the pathlength is computed from a known liquid absorbance using Beer's law in the form of Equations 4.1 and 4.2. Beer's law defines the relationship between the measured spectral signal intensity with respect to wavelength of incident energy with the concentration of a molecule as:

$$A = \varepsilon c l \tag{4.1}$$

Table 4.2 Optimum materials for ultraviolet window/cuvette materials.

Optical material	Transmittance range (nm)	Refractive index (at 600 nm)	Relative rupture strength (sapphire = 100)
UV-grade fused silica	200–2500	1.4580	10.9
Quartz, extremely low O-H	190–2500	1.5437	2.3
Sapphire (Al_2O_3)	150–5000	1.7677	100
Suprasil 300	190–3600	1.5400	3.8
Spectrosil	170–2700	1.5400	3.8
Infrasil	220–3800	1.5400	3.8

where ε is the molar absorptivity (referred to as molar extinction coefficient by earlier physicists) in units of Liter $\cdot$ Mole^{-1} $\cdot$ cm^{-1}; c is the concentration of molecules in the spectrometer beam in units of Mole $\cdot$ Liter^{-1} (Note: this is a scaled volume fraction unit); and pathlength (l) is the thickness of the sample in units of cm of the measured sample at a specific concentration. The pathlength for any specific molecule (or chemical compound) type is calculated by careful measurements of the absorbance (A) of a compound, generally diluted in a suitable organic solvent, and by applying the relationship:

$$l = \frac{A}{c\varepsilon} \tag{4.2}$$

For liquid analysis the sample type and special characteristics (i.e., material type, date opened, lot number, and supplier) should be noted. The pathlength may be precisely determined using this optical measurement technique when combined with a non-volatile pure organic liquid. Once the liquid is calibrated with a pure material, the pathlength of any cell or cuvette may be computed with a precise Absorbance measurement.

As a hypothetical example, one carefully mechanically measures 1 of 2 cuvettes with a caliper before filling them with pure hexane (86.18 g$\cdot$ mol^{-1}) at 20°C. For our example, the internal dimension of the measured cuvette is 1.0023 cm corresponding to an absorbance of 1.050 A (Au) at 300 nm. The second cuvette absorbance is measured to be 1.090 A. Any successive cuvette may have its pathlength calibrated simply by filling it with the pure hexane at 20°C and measuring its absorbance. The following relationship holds as Equation 4.3. Since one is calibrating for a single material, i.e. pure hexane at 20°C, then the molar concentration will always remain constant and the term may be dropped from the equation. The relevant terms then become:

$$\frac{A_n}{A_0} \cdot l_0 = l_n \tag{4.3}$$

where A_0 is the initial absorbance of the calibrated cuvette; A_n is the absorbance of a new cuvette; l_0 is the pathlength of the original cuvette; and l_n is the pathlength of the new cuvette. For the example given, the second cuvette has a pathlength as Equation 4.4.

$$\frac{1.090}{1.050} \cdot 1.0023\,\text{cm} = 1.0404\,\text{cm} \tag{4.4}$$

This is a valid method for calibrating pathlength noting that careful mechanical measurements are time consuming and may not be practicable for sealed cells and flow cells where the internal dimensions are not accessible. However once any

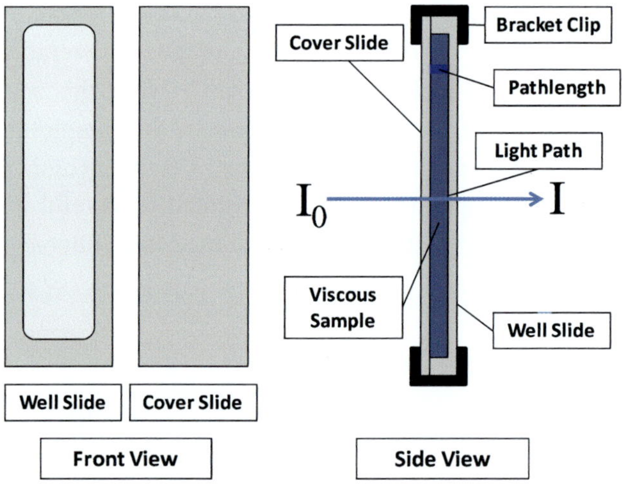

Fig. 4.3 Fixed pathlength well slide with holder for measuring slurries and viscous samples in transmittance mode.

cuvette is measured using the above method, any other cuvette pathlength may be similarly determined using a pure material and a spectroscopic measurement.

4.1.2.2 *Viscous Materials*

A simple and convenient method for measuring viscous materials using ultraviolet spectroscopy, is to utilize a well-slide with a specific pathlength in a transmittance measurement mode (Figure 4.3).

4.1.2.3 *Pastes, Emulsions, and Slurries*

Dependent on the constituents, pastes, and slurries generally may be handled in a similar manner as is proposed for viscous liquids. Always be aware of potential homogeneity problems. A set of well-slides (Figure 4.3) having varying pathlengths may be used for sample measurement, but if inhomogeneity is a problem, then the sample must be homogenized or carefully remixed to provide a more uniform suspension or emulsion prior to measurement.

With multiple phase emulsions or suspensions the analyst must take care that separation does not occur during the measurement process. At the window interface, the polar nature of the optical surface may cause segregation to occur. Common examples are emulsions, such as mayonnaise or margarine, and protein-containing materials; in this later case, the protein may actually form a coating on the optical window materials.

4.1.2.4 *Solids*

Solids, such as polymer samples, glass samples, thin layered samples, and fine ground samples (<1.0 mm diameter particles) are the typical solid samples most often measured using ultraviolet spectroscopy. These samples are generally measured "as received". In some cases the samples may be dried using a microwave oven or forced air drying methods to near a few percent retained water/moisture. The use of microwave drying carries with it the potential for damaging the sample due to high heat in various portions of the sample. Areas of particularly high heat occur due to the uneven field generation found in many commercial microwave heating systems.

In making spectroscopic measurements one must attempt to mitigate spectral artifacts caused by refractive index, scattering, and polarization effects. Samples must be carefully prepared so as to increase repeatability using a sufficient sample aliquot (i.e., sample volume) such that the measured sample is fully representative of the larger sample population. Solid polymer samples of sufficient thickness may also be used for direct transmittance or reflectance measurements using ultraviolet spectroscopy. Typically ultraviolet spectroscopy will penetrate many polymer samples from 0.1 to 2 mm or more in thickness, and one should be aware that as the ultraviolet energy penetrates the sample that an increase in thickness will cause greater intensity and also broadening of spectral bands.

(a) Powders

Unlike the many methods used to prepare powders for measurement using near infrared or infrared spectroscopy, for ultraviolet measurements the samples may be measured directly in reflectance or may be milled into particles using a cyclone-type mill with a 1.0 mm screen. This will make the sample of ground particles to be less than 1 mm in diameter. The powdered samples are placed in a sample cup specifically designed for dried and ground material. Such cups include the locking powder cup (Figure 4.4). The typical clear aperture of the window for sample measurement is from 20 mm to 25 mm. Sample may also be ground using a mortar and pestle that will produce a fine powder suitable for packing in a reflectance cup.

(b) Amorphous Materials (Organic), Continuous Sheets, and Polymeric Films

Polymer sheets may be measured in transmittance geometry or in diffuse reflectance geometry. A film holder may be used for transmittance measurements (Figure 4.5). For diffuse reflectance a standard reflectance cup with a diffuse back reflector may be used. The diffuse reflectance principle for film samples is illustrated in Figure 4.6. For this application a piece of the sheet or film is cut to the size of the sample cup and a diffuse reflectance measurement is made. It is important to remember that for a

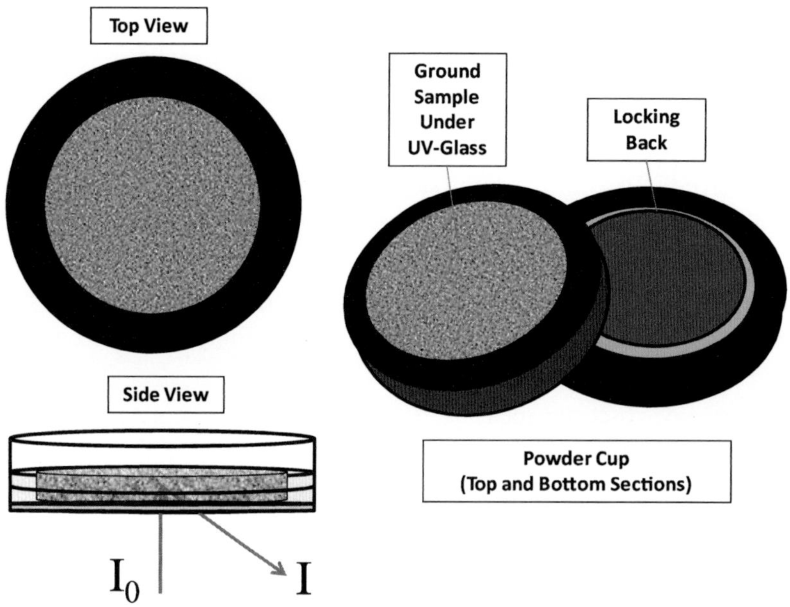

Fig. 4.4 Locking sample cup used for diffuse or specular geometry reflection measurements of powdered samples or other solids cut to conform to the sample holder.

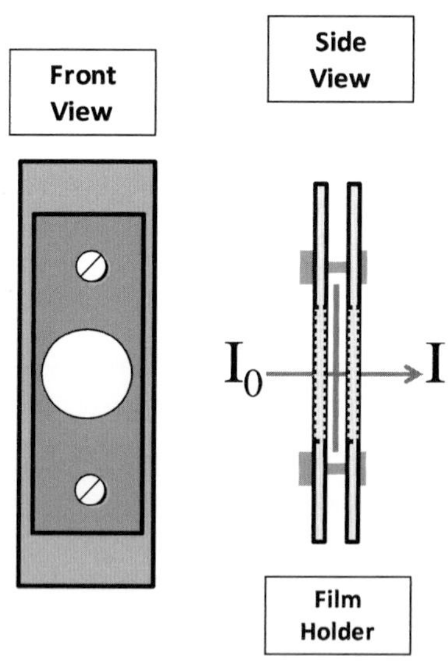

Fig. 4.5 Film holder for transmittance measurements of thin layers and films.

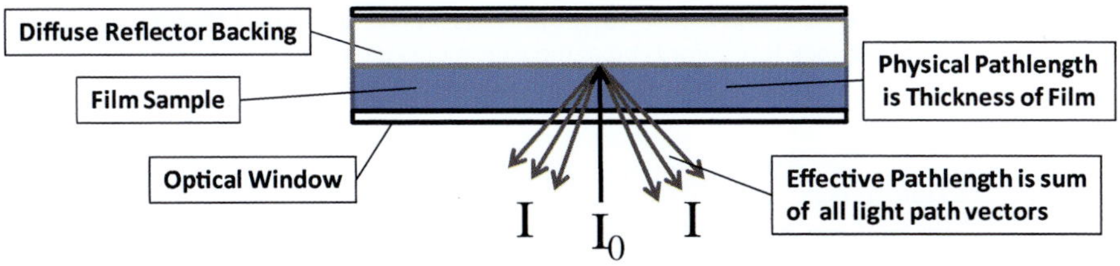

Fig. 4.6 Representation of a diffuse reflectance holder for thin film measurement.

sample with limited thickness the effective pathlength of a measurement is more than twice the physical pathlength of the sample. If the film sample has a metalized coating it will act as an etalon and give interference fringes in addition to the molecular spectrum. These fringes are due to multiple reflections within the film and have a periodicity proportional to the film thickness. See film thickness measurement in the index of this set of volumes. Other methods for handling polymers and films may be used from infrared sampling techniques, as demonstrated in Volume 4, Chapter 4. The reader is also referred to the index relating to the subjects of thin films and polymer sampling.

(c) Lumps, Granules, or Pellets

This particular class of solids is an extension of the sample types already discussed, and many of the procedures already highlighted may be used here. These materials may be measured in various diffuse reflectance cups in a bulk solids form, or as ground and mixed material for sample presentation using powder cups.

(d) Intractable Materials

Samples with high absorbance (or less than 1% reflectance) may not provide enough signal for reflectance measurements using many commercial ultraviolet spectrometers. If such samples must be measured routinely one may try a spectrometer with a much higher dynamic range. The measuring instrument must have at least one absorbance unit higher dynamic range than the typical absorbance for the sample(s) measured. Table 4.3 shows the minimum dynamic range for a spectrometer given a specified average sample absorbance.

4.1.2.5 *Sampling Methods*

This section provides similar information and cross-referenced information related to that found in the previous section of the chapter, but in this case it is more procedure oriented. Hints and tips are provided for the actual preparation method rather than focusing on the sample and its properties.

Table 4.3 Showing the relationship between %T and %R, T and R, and absorbance level. For typical measurements an instrument should have one (1.0) absorbance unit greater dynamic range than the sample.

%Transmittance or %Reflectance of sample	Transmittance or Reflectance of sample	Absorbance of sample	Minimum instrument dynamic range
100.0	1.0	0.0	1.0
10.0	0.1	1.0	2.0
1.0	0.01	2.0	3.0
0.1	0.001	3.0	4.0
0.01	0.0001	4.0	5.0
0.001	0.00001	5.0	6.0
0.0001	0.000001	6.0	7.0
0.00001	0.0000001	7.0	8.0

(a) Transmittance Windows

The most basic sampling device for ultraviolet spectroscopy includes a transmittance cell with a visible transmissive window; see Table 4.4 for a summary of common glasses and optical window materials.

(b) Transmittance Cells (Liquid Handling)

Liquids are generally measured using a fixed pathlength transmittance geometry with the majority of commercial ultraviolet instruments. Most instruments enable configuration for either reflectance or transmittance geometry, with the most common measurement being transmittance. Typical transmittance cells may be rectangular or rounded (cylindrical) in shape as illustrated in Figure 4.7.

Rectangular cuvets/cuvettes may be square shaped or be described as having an elongated rectangular cross-section with a variable pathlength (of approximately 1.0 mm to 20 mm), and with a width that fits into the sample holder of the instrument (i.e., generally 12.5 mm) (Figure 4.8). These cells are designed for ease of handling for sample measurements requiring increased pathlength.

(c) Gas Cells

Gas cells are considered a form of transmission/transmittance cells. Infrared is a technique often used for measuring gases due to the requirements for high resolution and high sensitivity. Ultraviolet is 100 to 10000 times less sensitive than infrared to the same molecular functional groups and thus ultraviolet is not considered to be useful as a laboratory or process gas measurement technique.

Table 4.4 Transmittance ranges for ultraviolet optical window and lens materials (in nm). The range from 200 nm to 400 nm is most useful for typical ultraviolet spectra.[6]

Optical material spectral ranges	Start	End
Silica, UV Grade Fused (SiO_2)	200	2500
Silica, Synthetic Fused (SiO_2)	230	2600
Quartz, Crystalline (SiO_2)	240	2600
Quartz, extremely low O-H	190	2700
Glass, Flint (SF 10)	380	2350
Glass, Flint (SF 8)	355	2350
Glass (BK 7)	315	2350
Glass, "Optical Crown"	320	2300
Glass, "Borosilicate Crown"	360	2350
Glass, Pyrex®	310	2500
Glass, Zerodur® Schott	400	2850
Glass, Tempax®	360	2350
Sapphire (Aluminum Oxide) (Al_2O_3) 5 mm is 50 %T at 5.1 microns	150	5100
Sodium Chloride (NaCl)	250	16000
Suprasil 300	190	3600
Diamond (C)	220	4000
Silica, Fused as pure SiO_2 (IR Quality)	200	4000
Quartz, Anhydrous	200	2750
Zirconium Fluoride (ZrF)	900	4760
Glass, Silica, Fused "Water Free" (SiO_2); 10 mm is 82%T at 3.15 microns	150	3150
Glass, Corning Optical Glasses as HPFS® 7978 and 7979 50%T/cm	150	4350
Glass, Corning Optical Glasses as HPFS® 7980; 50%T/cm	200	2600
Glass, Corning Optical Glasses as HPFS® 865X and 797X 50%T/cm	150	2600
Glass, ESCO Optical Glasses as S1-UVA and S1-UVB Working Range (Band 1)	170	2200
Glass, ESCO Optical Glasses as I2-IR Working Range	250	3600
Glass, ESCO Optical Glasses as G1 and A1 Working Range	270	2500
Glass, ESCO Optical Glasses as BK7 Working Range	350	2000
Ultraviolet Range (General Laboratory Use)	190	360

(d) Specular Reflectance Measurements

Specular reflectance (or mirror reflectance) represents an elastic photon reflection where only the direction of the incoming photon is changed, not its energy. Therefore the original energy (i.e., frequency or wavelength) is identical after the reflection. The most common applications of specular reflectance are for direct measurement of surfaces for contaminants or for measurement of optical properties of materials in the ultraviolet spectral region, for example, certain optical mirror or lens materials, paints or coatings. For the ultraviolet region, these measurements are referred to as transflectance or reflection/absorption. As may be observed in the Figure 4.9, for specular reflection measurements the incident angle and reflection measurement angle are the same. If the incident angle and the reflection measurement angle are

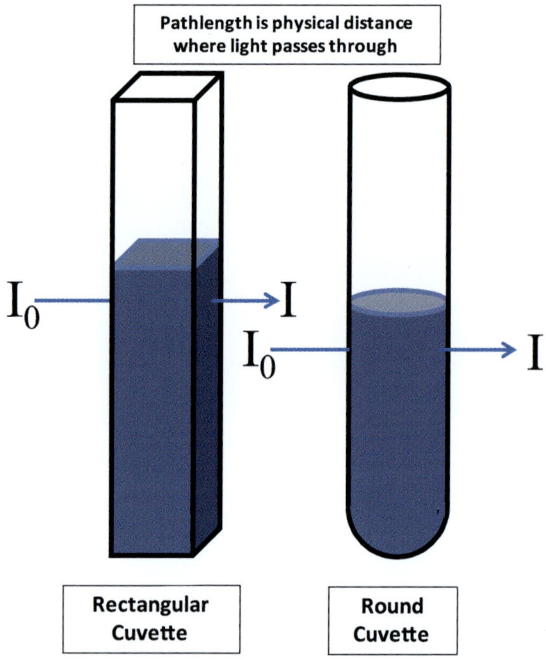

Fig. 4.7 Illustration of cylindrical and rectangular cuvet/cuvette types.

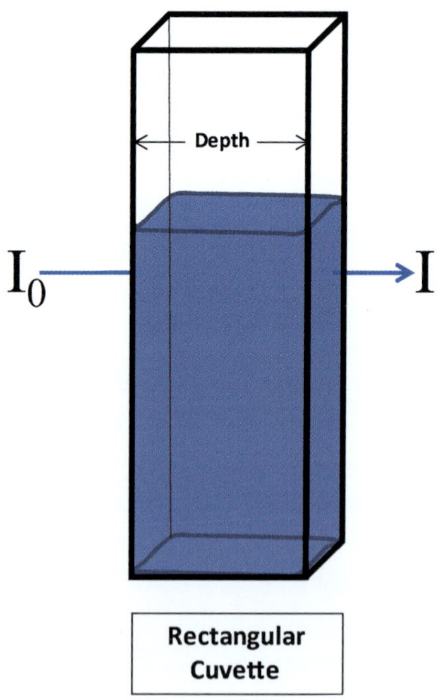

Fig. 4.8 Illustration of rectangular cuvet/cuvette with elongated pathlength.

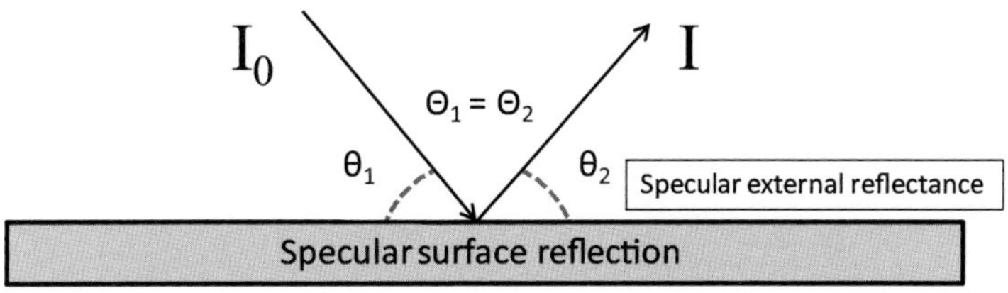

Fig. 4.9 Specular surface reflection/reflectance regimes.

different other reflective interactions will be measured, such as diffuse reflection, interactance, or transflectance. In these types of measurements an inelastic photon interaction occurs and an absorbance due to interaction with the sample is the most likely result.

(e) Other Sampling Techniques
Other sampling techniques used for ultraviolet spectroscopy, include polarized light, the integrating sphere, film thickness, refractive index of films, and fiber optics. These subjects are described in greater detail in other volumes and the main index for this set of volumes.

References

1. ASTM G130-12, Standard Test Method for Calibration of Narrow- and Broad-Band Ultraviolet Radiometers Using a Spectroradiometer, 2012. ASTM International, 100 Barr Harbor Drive, PO Box C700, West Conshohocken, PA, 19428-2959 USA.
2. ASTM STP500 (1972), Ultraviolet Spectroscopy, 1972. ASTM International, 100 Barr Harbor Drive, PO Box C700, West Conshohocken, PA, 19428-2959 USA.
3. ASTM E958-13, Standard Practice for Estimation of the Spectral Bandwidth of Ultraviolet-Visible Spectrophotometers, 2013. ASTM International, 100 Barr Harbor Drive, PO Box C700, West Conshohocken, PA, 19428-2959 USA.
4. ISO-21348, ISO 21348 Definitions of Solar Irradiance Spectral Categories, ISO 21348 information is provided courtesy of Space Environment Technologies spacenvironment@spacenvironment.net http://SpaceWx.com.
5. W. Tobiska and A. Nusinov, ISO 21348-Process for determining solar irradiances, In *36th COSPAR Scientific Assembly*, 36 (200^), 2621.
6. J. Workman, *The Academic Press Handbook of Organic Compounds: visible , IR, Raman, and UV-VIS Spectra Featuring Polymers, and Surfactants.* 3 Volume Set: Vol. 1, Methods and Interpretation; Vol. 2, UV-Vis and visible Spectra; Vol. 3, IR and Raman Spectra. (Academic Press, Boston, 2000).
7. J. J. Workman and A.W. Springsteen (Eds.), *Applied Spectroscopy: A Compact Reference for Practitioners* (Academic Press, Boston, 1998).
8. J. R. Lakowicz, *Principles of Fluorescence Spectroscopy*, 3rd edn. (Springer Science & Business Media, 2011).

9. P. Misra and M.A. Dubinskii (Eds.), *Ultraviolet Spectroscopy and UV lasers*, 1st edn. (Taylor and Francis, CRC Press, 2002).
10. J. D. Ingle Jr and S.R. Crouch, *Spectrochemical Analysis*, 1st edn. (Prentice Hall, 1988).
11. H.-H. Perkampus, *UV-VIS Spectroscopy and its Applications,* 1st edn. (Springer-Verlag, Berlin, 1992).

Chapter 5

Data Acquisition for Ultraviolet Spectra

Contents

5.1 Recommended Spectrum Acquisition, Presentation Format and Data Manipulation Procedures

5.1.1 *Recommended Spectrum Acquisition Parameters*

The following are data acquisition guidelines for average ultraviolet measurement scenarios. A standard transmittance measurement using a typical ultraviolet spectrophotometer will complete a single replicate scan in from 0.1 seconds to 5 seconds. A complete set of replicate scans giving sufficient signal to noise for most applications will require from 15 to 60 seconds (typical), with spectral resolution of approximately 1 nm bandwidth/halfwidth/linewidth. This measurement time provides a signal-to-noise ratio (S/N) of between 1000:1 and 3000:1. If a spectrum is to be used for qualitative purposes only, then a S/N of approximately 1500:1 is normally sufficient and is typically measured in under 30 seconds (total data acquisition time).

For quantitative applications, or applications involving spectral subtraction, higher signal-to-noise performance is required at better than 2000:1. Note that increased replicate scanning will at some point approach a point diminishing returns in terms of spectral quality (i.e., low noise characteristics) and S/N improvement. Signal-to-noise improvement is limited by the square root of *n* law for random variation, or white noise, where n is the number of replicate scans. Ultraviolet spectrophotometers may drift over a several minute period and structured noise, due to temperature changes, may become a part of spectra measured for extended periods. Note that for ultraviolet work a halfwidth of greater than 1 nm has no real advantage as spectral bands are somewhat broader than this.

Irrespective of the final selection of data acquisition parameters, full documentation of the experimental details should be recorded with all spectra, especially if regulated or audited procedures are followed. Also, as a general rule, it is recommended that good laboratory practice guidelines are adopted throughout, which include laboratory notebooks, either paper or electronic and data recorded to include measurement data (sample and background), measurement parameters, analyst, results, and corroborating signatures.[1]

5.1.2 *Recommended Spectral Range*

Unless special circumstances dictate otherwise, visible spectra are recorded from a minimum of 190 nm to 400 nm, however most UV-Vis spectrophotometers have a scan range of from 190 nm to 1100 nm. As with other spectrophotometers, ultraviolet instruments are restricted by detector used, source used, and functional optics. Diode array instruments are limited by the array detector sensitivity. With modern rapid scanning ultraviolet spectroscopy instruments, it is normal to measure the entire spectral range with each measurement (i.e., 190 nm to 1100 nm), and at full data resolution (i.e., greatest number of data points). When calibration or data transfer is required then data may be truncated (or windowed) so that one spectrometer collects the identical spectral range and data points as another spectrometer. Such data management allows easier compatibility of databases and calibrations developed using regression methods for quantitative calibration. Note: information on calibration transfer methods and data collection methods for calibrating instruments may be found within this volume index or the main index, or within Chapter 6 of each volume, covering data analysis.

5.1.3 *Recommended Presentation Formats*

The normal intensity scale obtained for a raw measurement is %R or %T, usually scaled from 0 to 100%. If an abbreviated or expanded scale is used, it is important to ensure that the scale limits are presented, preferably with immediate scale markings shown. Note that for all numerical operations on spectral data, it is necessary to perform the operations on the absorbance form of the spectrum, including baseline corrections, spectral subtractions, and any numerical scaling of data. An important exception to this is with the numerical and/or digital treatment of noise. Noise is essentially linear over the entire scale in reflectance (%R), or transmittance (%T) and is logarithmic in the absorbance scale. All processing that assumes a linear distribution of noise, such as digital smoothing, and digital signal processing, is preferably applied to the reflectance or transmittance form of the data in order to perform a proper correction. After this correction the data may be transformed to absorbance scale using Beer's law computations, described in many sections throughout this set of volumes. For example, see Front Material and Chapter 2, Section 2.1.3.

The standard linear wavelength scale is normally used for ultraviolet data, however, in physics it is entirely appropriate to represent ultraviolet spectra in frequency or wavenumber space as well.

5.1.4 *Spectral Data Manipulation*

As in other forms of electronic and molecular spectroscopy, ultraviolet absorbance spectra are recorded using ratioed spectral data. This ratioed data consists of the sample spectrum, the reference spectrum, and the dark signal spectrum. The use of Beer's law for spectroscopic measurements is shown in the following description. To convert the light transmitted from a sample (I) ratioed to the incident energy (I_0) to Absorbance (A), as a linear estimate of spectral response to concentration, the following equations are used. If only an internal reference is used for spectral collection, then the absorbance spectrum with respect to wavelength is computed as:

$$\mathrm{A} = \log_{10}\left(\frac{I_0}{I}\right) = \log_{10}\left(\frac{1}{\mathrm{T}}\right) = \log_{10}\left[\left(\frac{\mathrm{R_I} - \mathrm{D_{R_I}}}{\mathrm{S} - \mathrm{D_S}}\right)\right] \tag{5.1}$$

If an external standard material is used to calibrate the internal reference material, then the absorbance spectrum with respect to wavelength is computed as:

$$A = \log_{10}\left(\frac{I_0}{I}\right) = \log_{10}\left[\left(\frac{\mathrm{R_I} + \mathrm{R_\Delta} - \mathrm{D_{R_I}}}{\mathrm{S} - \mathrm{D_S}}\right)\right] \tag{5.2}$$

where:

$$R_\Delta = \frac{(R_E - D_{RE})}{(R_I - D_{RI})} \tag{5.3}$$

and where S is the sample measurement; D_S is the Dark measurement for the Sample; R_I is the internal Reference Measurement; D_{RI} is the Dark Measurement for the Internal Reference Sample; R_E is the measurement of the external standard reference material for calibration; D_{RE} is the Dark Measurement for the external standard reference material; and R_Δ is the correction factor between the dark corrected R_E and R_I. The final spectrum is simplified to the ratio of (Internal Reference corrected by External Reference minus Dark) divided by (Sample-Dark). Note that Dark is a measurement where no energy from the source is allowed to the detector (it is blocked). The Dark measurement represents the dark current changes in the instrument during a measurement that relate to the electronic noise where the detection electronics, computational electronics, and control electronics perform basic measurement functions with no imputed energy reaching the detector (i.e., the source is turned off or blocked using a shutter closure during the Dark spectrum measurement).

For spectral appearance purposes, digital smoothing and nominal baseline corrections are acceptable if they improve overall quality of the final spectra for qualitative evaluation or quantitative analysis. If there are any doubts in terms of acceptability for data processing, the original raw data should be archived and properly labeled as to the exact method and sampling procedure used to prepare the sample and measure the data. All documentation concerning data manipulation should be retained with the spectral file in a digital format.

5.1.5 *Data Formatting*

A standard data format is essential for the utilization of ultraviolet spectral data. Data from various instruments is routinely combined for quantitative calibration transfer and utilization of spectral libraries for qualitative searches for the purpose of identification.

The goal in measuring ultraviolet data is the ability to produce an exact and very reproducible standard spectrum for any given material using a prescribed method of sample preparation, independent of operator or instrument (model or vendor). Many of the factors that impact meeting that goal are known, and it is anticipated that future generations of instrumentation, with integrated sampling and data manipulation, may lead to continuous improvements in data measuring and archiving. Readers are encouraged to utilize the standard procedures for sample preparation as defined in texts, such as the ASTM International *Annual Book of Standard's.*[2] Each instrument manufacturer may have from 1–10 (or more) data formats including older versions of their software. A list of the most typical spectroscopy software data formats is given in Table 5.1.

Table 5.1 Typical spectroscopic data file formats and extensions.

Data format name	Extension
AB SCIEX Data Explorer	(.pkm)
ASCII Table	(.csv)
Bruker	(.XML)
Excel	(.xls)
Finnegan	(.asc)
JCAMP-DX	(.dx)
MatLab	(.mat)
mzData	(.XML)
Perkin Elmer	(.sp)
Plain Text	(.txt)
Spectra-Calc (Archival)	(fixed point .spc)
Thermo Galactic	(.spc)
XLM Marked-up text	(.XML)

5.1.6 *Summary*

This chapter has focused on the basic issues of routine sampling for qualitative and quantitative ultraviolet spectra. There are many routine specialized approaches to sampling that may be adopted for research applications. These are typically experiment dependent, and their success may be somewhat limited to the application for which they are customized.

References

1. Guidance for Industry Good Laboratory Practices Questions and Answers, U.S. Department of Health and Human Services Food and Drug administration Office of Regulatory affairs, June 1981 (Minor editorial and formatting changes made December 1999 & July 2007).
2. ASTM (American Society for Testing and Materials) Method E169-04 (2009). Standard Practices for General Techniques of Ultraviolet-Visible Quantitative Analysis, ASTM Committee E 13, 1916 Race Street, Philadelphia, PA 19103.

Chapter 6

Data Analysis and Presentation for Ultraviolet Spectroscopy

Contents

This chapter discusses the mathematical data treatments common to ultraviolet spectroscopy. It is intended to be a reference chapter for descriptions and explanations of each type of mathematical function common to ultraviolet methods. Chemometrics, multivariate regression, and statistical methods are covered in other volumes of this set. The reader is referred to the master index for individual topics related to specific mathematical data treatments. Topics such as matrix mathematics, regression, partial least squares, quantitative analysis, qualitative analysis, and other regression and calibration related topics are included in other volumes.

6.1 Baseline Correction Algorithms

Baseline correction algorithms are used to correct curvature or offset of the measured spectrum signal due to particular physical or chemical properties. Scattering due to small particle size or mixtures of scattering and absorbing materials may complicate the baseline curvature or offset effects. Additional details describing baseline correction may be found in the main index.

6.1.1 *Baseline Offset Correction*

Offset correction is performed by selecting a single point or multiple points along a spectrum and adding or subtracting a y-value (i.e., intensity value) at each selected point or points to correct or shift the baseline offset. This preprocessing step is used to align the baseline of two or more spectra, causing them to overlap at a predetermined baseline intensity value (e.g., 0 absorbance). Any baseline may be corrected to a single or multiple points that are set to specific predetermined offset values. This simple process is also used to bring the minimum point to zero absorbance whenever desired. A single point offset spectrum for −0.05 A and +0.05 A is shown for all wavelengths, as compared to the original spectrum, in Figure 6.1.

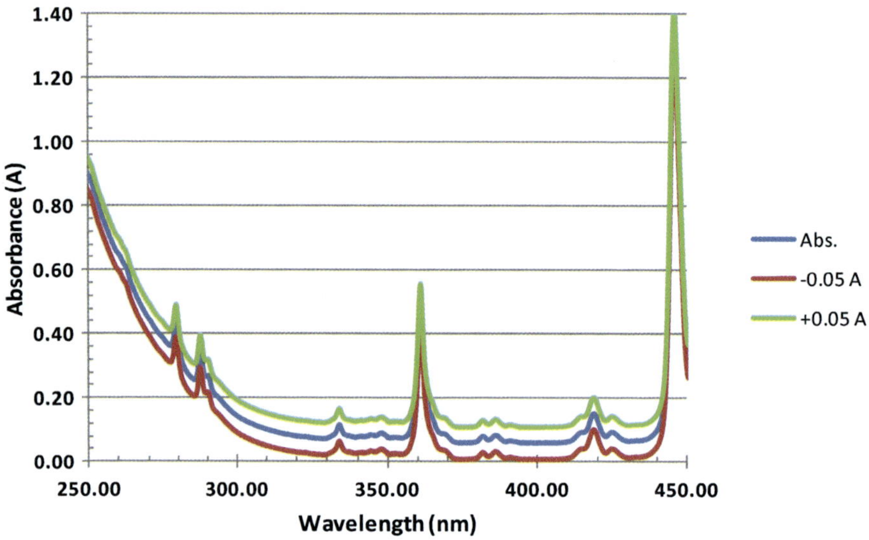

Fig. 6.1 Single point baseline offset of −0.05 and +0.05 Absorbance (A) units for a Holmium Oxide spectrum from 250 to 450 nm.

6.1.2 *Derivatives*

Derivatives are used to remove offset or some slope characteristics in the baseline which are the result of background differences or scattering effects. The reader is referred to the index for a detailed section on derivative methods and spectroscopy. Derivatives are demonstrated within these volumes in significant detail, please refer to the master index.

6.1.3 *Flat Baseline Correction (Two-Point Linear Baseline Correction)*

For this correction, two data points are selected on the spectrum signal, and a line connecting these points is subtracted from the signal. This process removes first-order uniform offset and will remove some slope effect.

This form of correction subtracts a regression line between two (2) selected data points on the spectrum and removes the slope or linear portion of a curved baseline from the spectrum. This form of baseline correction uses 2 selected data points and the equation as follows:

$$y = a + bx \text{ (for 2 data points)} \tag{6.1}$$

This linear baseline equation has the form using a as the intercept and b as the slope or regression coefficient. From Figure 6.2 one observes that the line formed using the 2 points (in red) is subtracted from the original curved spectrum (blue),

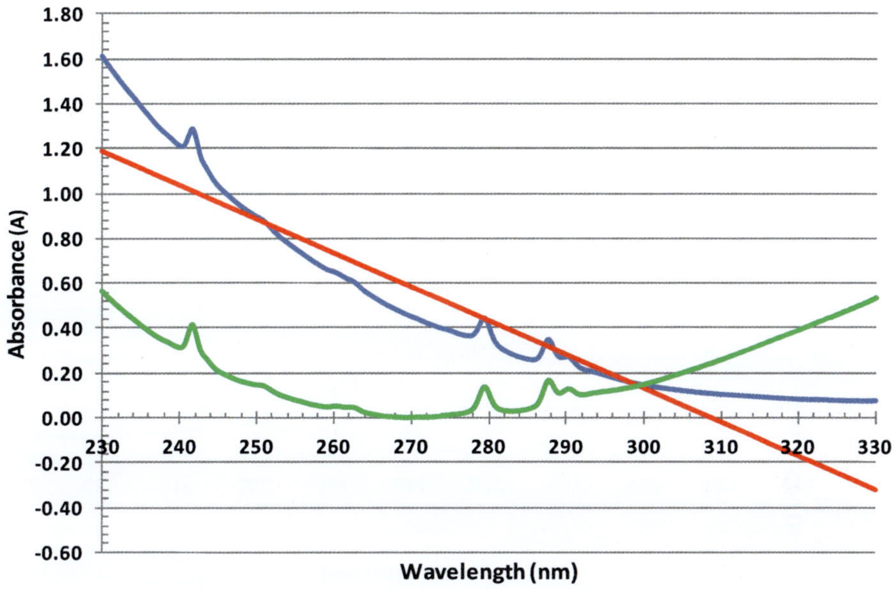

Fig. 6.2 This figure shows a 2-point linear baseline correction. One observes the original, curved, uncorrected spectrum (blue), the fitted regression line to the data at 250 and 300 nm (manually selected data points) shown in red, and the corrected spectrum after subtraction of the 2 point regression line and addition of the offset adjustment to correct the minimum absorbance to 0.0 (green).

and the minimum point is moved to 0.0 absorbance by adding an offset to align the minimum absorbance to 0.0. This final corrected spectrum is shown in green.

The process then is to select 2 data points, compute the regression line, subtract the regression line from the original spectrum, and add an offset to bring the minimum absorbance to 0.0.

6.1.4 *Linear Baseline Correction (Best Fit to All Points)*

All data points are regressed using a simple linear regression for the spectrum signal, and a line fitting these points is subtracted from the signal. This process removes first-order uniform offset and will remove some slope effect more accurately for a severely curved spectrum than two (or a few) selected data points.

This form of baseline correction subtracts the regression line computed for all data points on the spectrum and removes the linear portion of a curved baseline from the spectrum. This form of baseline correction uses all data points and the equation as:

$$y = a + bx \text{ (for all data points)} \tag{6.2}$$

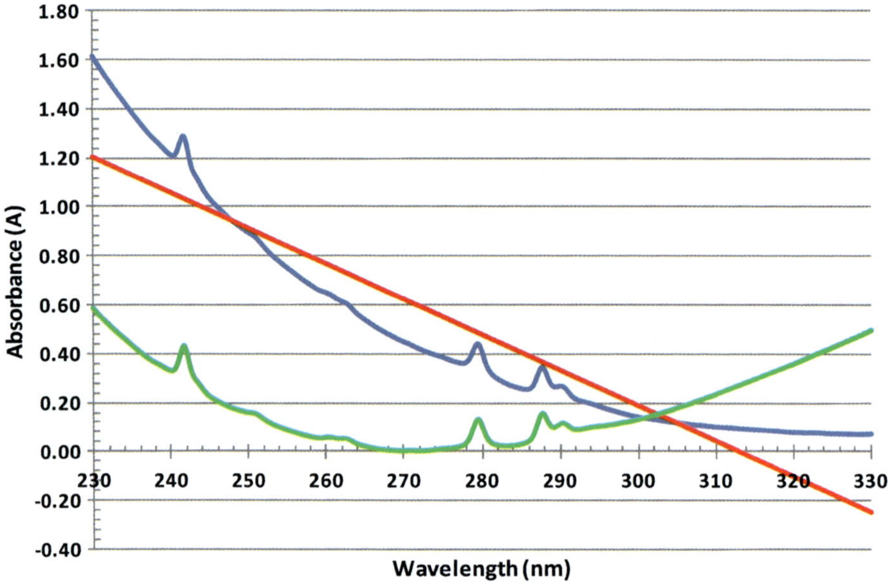

Fig. 6.3 This figure shows an all data point linear baseline correction. One observes the original, curved, uncorrected spectrum (blue), the fitted regression line to all the data points (red), and the corrected spectrum after subtraction of the full regression line and addition of the offset adjustment to correct the minimum absorbance to 0.0 (green).

This linear equation has the form using a as the intercept and b as the slope or regression coefficient. From Figure 6.3 one observes that the line formed using all data points (in red) is subtracted from the original curved spectrum (blue), and the minimum point is moved to 0.0 absorbance by adding an offset to align the minimum absorbance to 0.0. This final corrected spectrum is shown in green.

The process then is to use all spectral data points, compute the regression line, subtract the regression line from the original spectrum, and add an offset to bring the minimum absorbance to 0.0.

6.1.5 *Polynomial or Baseline Fitted Correction*

For this baseline correction approach, all data points of the spectrum are regressed using a polynomial regression for the spectrum signal, and a curve fitting these points is subtracted from the signal. This process removes higher-order curvature, baseline offset, and some slope effect more accurately than a linear baseline correction.

This form of correction subtracts the non-linear polynomial regression curve computed for all data points on the spectrum and removes the fitted curvature portion of the curved baseline from the spectrum. This form of baseline correction uses all

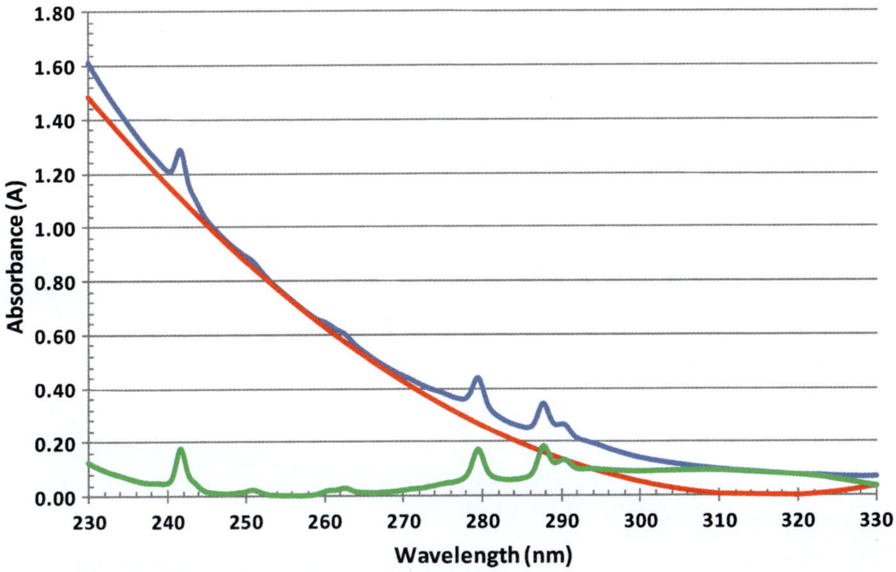

Fig. 6.4 This figure shows an all data point polynomial baseline correction. One observes the original, curved, uncorrected spectrum (blue), the fitted polynomial regression to all the data points (red), and the corrected spectrum after subtraction of the full polynomial regression and addition of the offset adjustment to correct the minimum absorbance to 0.0 (green).

data points and the Equation 6.3:

$$y = b_n x^n + b_{n-1} x^{n-1} + a \text{ (for all data points)} \tag{6.3}$$

This polynomial regression equation has the form using a as the intercept term and b's as the slope or regression coefficient terms. From Figure 6.4, one observes that the polynomial curve formed using all data points (in red) is subtracted from the original curved spectrum (blue), and the minimum point is moved to 0.0 absorbance by adding an offset to align the minimum absorbance to 0.0. This final corrected spectrum is shown in green.

The process then is to use all spectral data points, compute the polynomial regression, subtract the polynomial regression from the original spectrum, and add an offset to bring the minimum absorbance to 0.0.

6.1.6 *Exponential Fitted Baseline Correction*

For the exponential fitted baseline correction, all data points are regressed using an exponential regression for the spectrum signal, and a curve fitting these points is subtracted from the signal. This process removes higher-order curvature, baseline offset, and some slope effect more accurately than a linear baseline correction.

This form of correction subtracts the non-linear exponential regression fitted curve computed for all data points on the spectrum and removes the fitted exponential curvature portion of the curved baseline from the spectrum. This form of baseline correction uses all data points and the equation as follows:

$$y = ae^{bx} \text{ (for all data points)} \tag{6.4}$$

This exponential regression equation has the form using $a > 0$ and b as the slope or regression coefficient terms, and e is Euler's number (constant) and is the base of the natural logarithm or 2.71828.

From Figure 6.5 one observes that the exponential curve formed using all data points (in red) is subtracted from the original curved spectrum (blue), and the minimum point is moved to 0.0 absorbance by adding an offset to align the minimum absorbance to 0.0. This final corrected spectrum is shown in green.

The process then is to use all spectral data points, compute the exponential regression, subtract the exponential regression from the original spectrum, and add an offset to bring the minimum absorbance to 0.0.

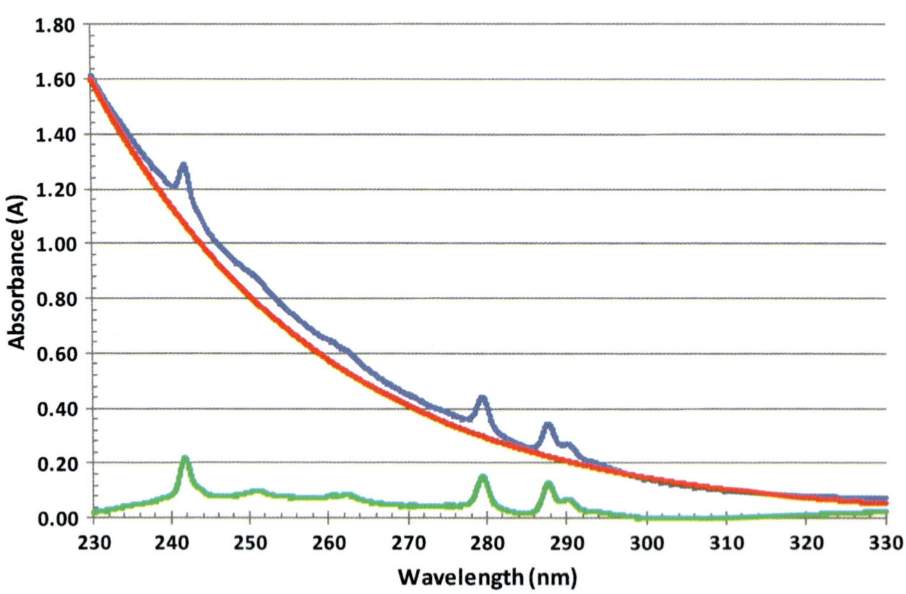

Fig. 6.5 This figure shows an all data point exponential baseline correction. One observes the original, curved, uncorrected spectrum (blue), the fitted exponential regression fitted to all the data points (red), and the corrected spectrum after subtraction of the full exponential regression and addition of the offset adjustment to correct the minimum absorbance to 0.0 (green).

6.1.7 *Power Fitted Baseline Correction*

All data points are regressed using a power function regression for the spectrum signal, and a curve fitting these points is subtracted from the signal. This process removes higher-order curvature, baseline offset, and some slope effect, but with poorer results on highly curved spectra as compared to polynomial and exponential fitted functions.

This form of correction subtracts the non-linear power function regression fitted curve computed for all data points in the spectrum and removes the fitted power function curvature portion of the curved baseline from the spectrum. This form of baseline correction uses all data points and the equation as:

$$y = ax^b \text{ (for all data points)} \tag{6.5}$$

This power function regression equation has the form using $a > 0$ and b as the slope or regression coefficient terms.

From Figure 6.6 one observes that the power function curve formed using all data points (in red) is subtracted from the original curved spectrum (blue), and the

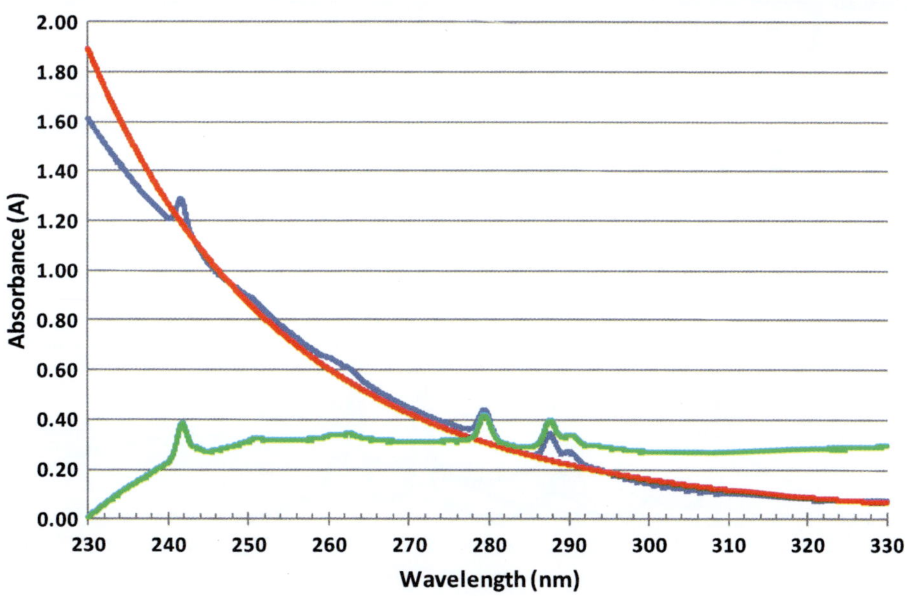

Fig. 6.6 This figure shows an all data point power function baseline correction. One observes the original, curved, uncorrected spectrum (blue), the fitted power function regression fitted to all the data points (red), and the corrected spectrum after subtraction of the full power function regression and addition of the offset adjustment to correct the minimum absorbance to 0.0 (green).

minimum point is moved to 0.0 absorbance by adding an offset to align the minimum absorbance to 0.0. This final corrected spectrum is shown in green.

The process then is to use all spectral data points, compute the power function regression, subtract the power function regression from the original spectrum, and add an offset to bring the minimum absorbance to 0.0.

6.1.8 *Comparison of Methods*

A comparison of the corrected spectra for these baseline methods is shown in Figure 6.7 and data displayed in Table 6.1. From the table, it is observed that for this highly curved ultraviolet spectrum that the polynomial baseline corrected spectrum was slightly improved as compared to the exponential, and the linear fitted corrections were the worst. This is indicated by the highest to lowest coefficient of determination (R-Squared, or R^2) for the fitted functions. This procedure may be used for finding the best fit correction approach from a suite of baseline correction algorithms.

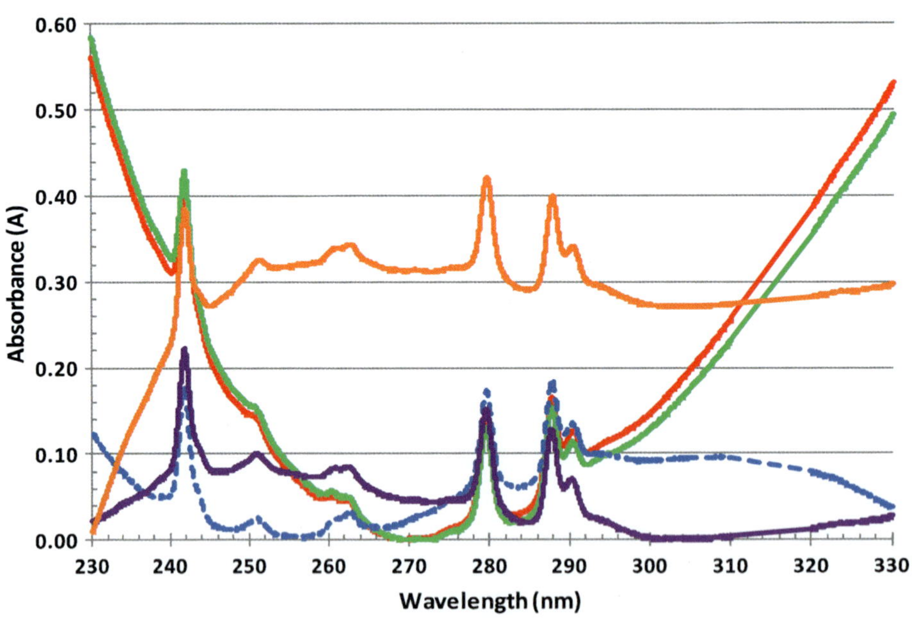

Fig. 6.7 A comparison of the corrected curved ultraviolet spectrum. The 2-point linear (red), all-point linear (green), polynomial (blue, dotted), exponential (violet), and power function (orange) corrections are shown. As demonstrated in the figure and Table 6.1, the polynomial is slightly better for this specific spectrum, based on the best fitted function with the highest R-squared value.

Table 6.1 Showing R-squared values for each fitted function.

Correction method	Data points	R-squared
2 Point Linear	2	NA
Linear	All	0.8656
Polynomial	All	0.9920
Exponential	All	0.9907
Power function	All	0.9876

Table 6.2 Hypothetical linear data used for illustrations of section 6.2 data analysis.

Absorbance	Reference values	Linear prediction	Error	Polynomial prediction	Error
0.18	2.00	2.10	−0.10	2.08	−0.08
0.36	4.00	3.92	0.08	3.92	0.08
0.56	6.00	5.95	0.05	5.96	0.04
0.75	8.00	7.88	0.12	7.89	0.11
0.97	10.00	10.12	−0.12	10.12	−0.12
1.17	12.00	12.15	−0.15	12.15	−0.15
1.34	14.00	13.88	0.12	13.87	0.13
AVERAGE			0.00		0.00
STDEV			0.12		0.12
R^2			0.9993		0.9993
r			0.9996		0.9996

6.2 Data Analysis: Plotting Data for Visual Assessment

The following discussion refers to hypothetical data as given in Tables 6.2 and 6.3 for linear and curved data sets, respectively. The tables demonstrate both linear and polynomial regression methods used for the specific data types. The table columns include Absorbance data, Reference Values, Linear Estimation (Prediction) Values, Error, Polynomial Estimation (Prediction) Values, and Error. Summary statistical parameters are also shown for both the linear and polynomial regression methods, respectively.

The spectroscopist is able to visually identify spectral shapes and differences simply by observing various types of data plots. This section describes the techniques used for manual observation of spectral data to determine its characteristics. There are various plotting techniques for spectral data that demonstrate its character,

Table 6.3 Hypothetical curved data used for illustrations of section 6.2 data analysis.

Absorbance	Reference values	Linear prediction	Error	Polynomial prediction	Error
0.15	2.00	3.86	−1.86	2.19	−0.19
0.25	4.00	4.84	−0.84	4.21	−0.21
0.35	6.00	5.81	0.19	6.03	−0.03
0.43	8.00	6.59	1.41	7.35	0.65
0.60	10.00	8.25	1.75	9.76	0.24
0.90	12.00	11.18	0.82	12.66	−0.66
1.34	14.00	15.47	−1.47	13.81	0.19
AVERAGE			0.00		0.00
STDEV			1.42		0.41
R^2			0.8922		0.9908
r			0.9446		0.9954

features and the relationships between variables. Several examples are illustrated in this section.

6.2.1 *Using Two-Dimensional Data Plots for Visual Data Interpretation*

The main uses of residual plots are to provide a simple tool for interpretation of the quality or accuracy of predictive calibration models. The basic residual plots used for quantitative analysis include:

- Predicted vs. Actual
- Residual vs. Actual

6.2.1.1 *Actual Values versus Predicted Values Plot*

The predicted versus actual plot is made by plotting the reference or actual values (e.g., mass per unit volume, mg/L) for each sample as the abscissa (X axis) versus the spectroscopic method (predicted value) as the ordinate (Y axis). This X-Y plot demonstrates the basic correlation between the estimated value using the analytical method and the actual reference values (or primary reference chemistry) as a comparison. It also indicates precision and fitness to the actual (reference) values. A linear data set with a linear regression line fit is shown in Figure 6.8.

6.2.1.2 *Actual versus Residual Plot*

The actual (or primary reference) value minus the predicted (spectroscopic estimated value) is equal to the residual value for each calibration sample. The actual (reference value) versus the residual value plot is a powerful visual tool for diagnosing

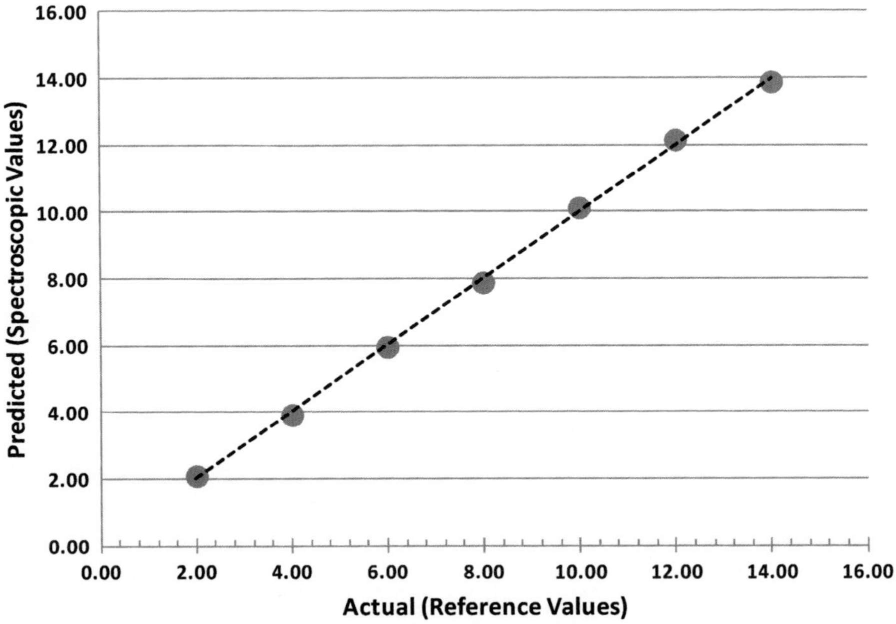

Fig. 6.8 Predicted versus actual plot for Linear data in Table 6.2. A strong positive correlation is shown with good fit for predicted values versus reference values.

issues associated with evaluating quantitative methods for spectroscopy. To generate this plot, one plots the actual or reference value as the abscissa (X axis) and the difference between the spectroscopic method and the reference method as the ordinate value (Y axis); this is plotted for each test or calibration/prediction sample. Figure 6.9 demonstrates the basic prediction residuals versus the reference results for a prediction set representing Table 6.2 for a linear fit regression. Note there are no slope or bias differences between the reference values and the predicted values.

6.2.2 *Using Regression Plots to Fit Linear and Curved Data Types*

Various forms of regression are useful for construction of calibration equations for the purpose of using measured absorbance data to predict or estimate concentration data. A set of samples is constructed with known concentrations, the absorbance values are measured, and a plot is constructed such as shown in Figure 6.10. A regression fit is constructed to this X-Y data such that an absorbance measurement of an unknown (or test) sample may be used in the regression equation to estimate (or predict) the concentration for that sample. Various linear and non-linear regression methods may be applied. The following section described the use

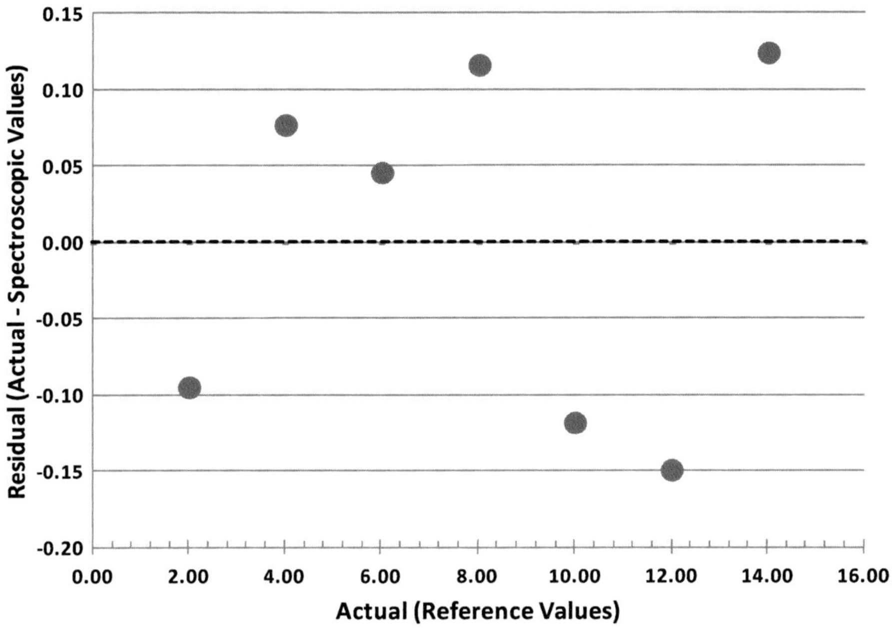

Fig. 6.9 Actual versus residual plot for linear data in Table 6.2. A strong positive correlation is indicated with excellent (precision), no slope or bias differences are shown; no curvature to data for residual values versus reference values is demonstrated.

of linear and polynomial regression to fit linear and curved (non-linear) data sets, respectively.

6.2.2.1 *Linear Regression on Linear Data*

Table 6.2 Linear regression on linear data is illustrated in Figure 6.10. The regression line closely fits the calibration data and this is indicated by a standard deviation of difference between the regression estimate and the actual values of 0.12 and a coefficient of determination (as R-square) of 0.9993.

6.2.2.2 *Polynomial Regression on Linear Data*

The Table 6.2 Polynomial regression on linear data is illustrated in Figure 6.11. The regression also closely fits the calibration data and this is indicated by a standard deviation of difference between the regression estimate and the actual values of 0.12 and a coefficient of determination (as R-square) of 0.9993; this is precisely the same result as the linear fit to the data. A comparison of the two methods is shown as a residual plot in Figure 6.12.

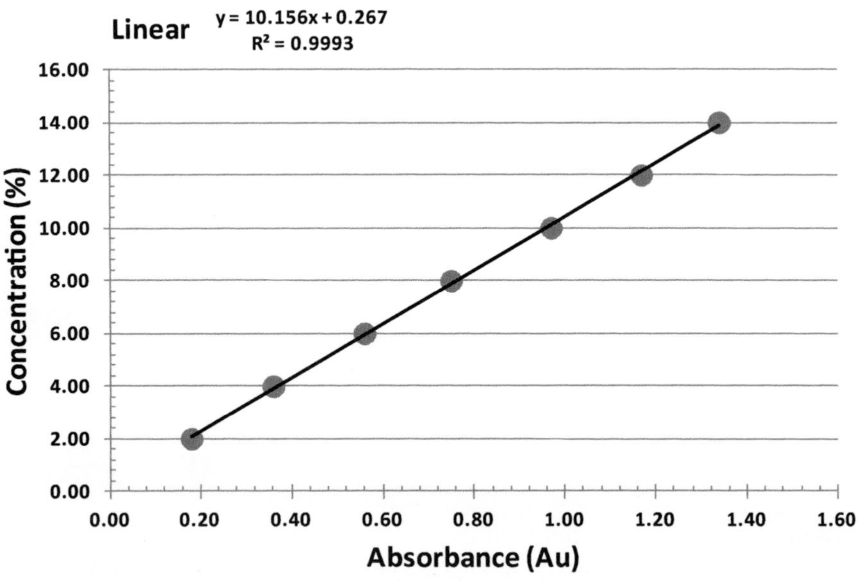

Fig. 6.10 Absorbance versus actual (reference) concentration data plot for a linear regression fit to the linear data in Table 6.2. A strong positive correlation is shown with good fit for absorbance values versus reference concentration values.

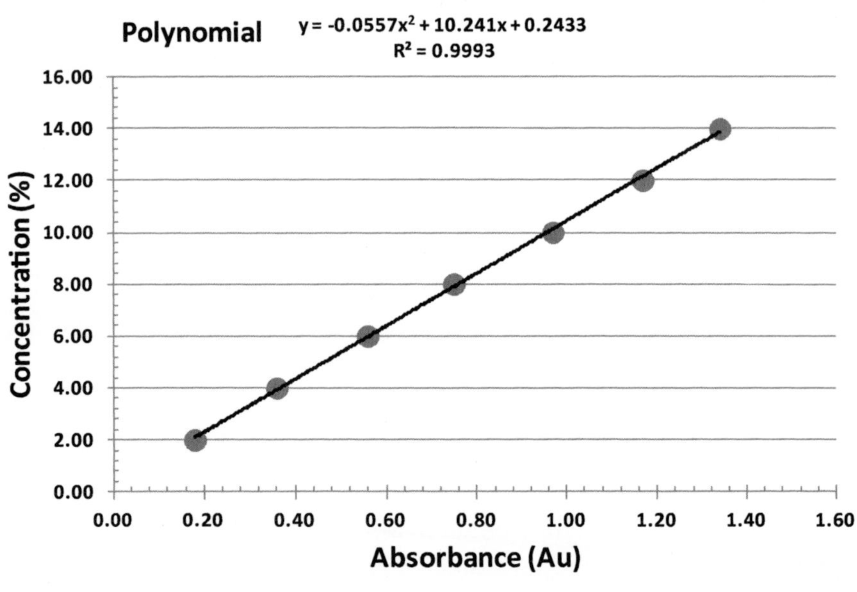

Fig. 6.11 Absorbance versus actual (reference) concentration data plot for a polynomial regression fit to the linear data in Table 6.2. A strong positive correlation is shown with good fit for absorbance values versus reference concentration values. This result indicates that a polynomial fit may be recommended as a universal regression method.

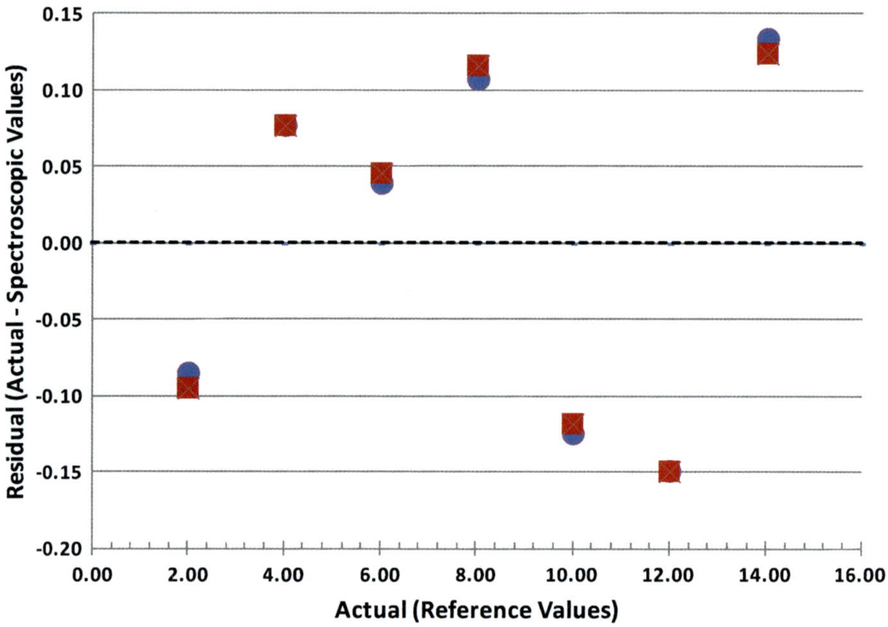

Fig. 6.12 Actual versus residual plot for linear and polynomial regression fitted equations for linear data from Table 6.2. A strong positive correlation is indicated with excellent (precision), no slope or bias differences are shown; no curvature to data for residual values versus reference values is demonstrated. The linear regression and polynomial regression yield nearly identical fitted results.(Linear fit residuals are brown squares, polynomial fit residuals are blue circles).

6.2.2.3 *Linear Regression on Curved (Non-linear) Data*

The Table 6.3 linear regression on non-linear or curved data is illustrated in Figure 6.13. As anticipated, the linear regression does not fit the curved calibration data and this is indicated by a standard deviation of difference between the regression estimate and the actual values of 1.42 and a coefficient of determination (as R-square) of 0.8922; this is not an acceptable calibration model or equation and is not suitable for prediction use.

6.2.2.4 *Polynomial Regression on Curved (Non-linear) Data*

The Table 6.3 polynomial regression on non-linear or curved data is illustrated in Figure 6.14. As anticipated, the polynomial regression does fit the curved calibration data quite well and this is indicated by a standard deviation of difference between the regression estimate and the actual values of 0.41 and a coefficient of determination (as R-square) of 0.9908; this is an acceptable calibration model or equation.

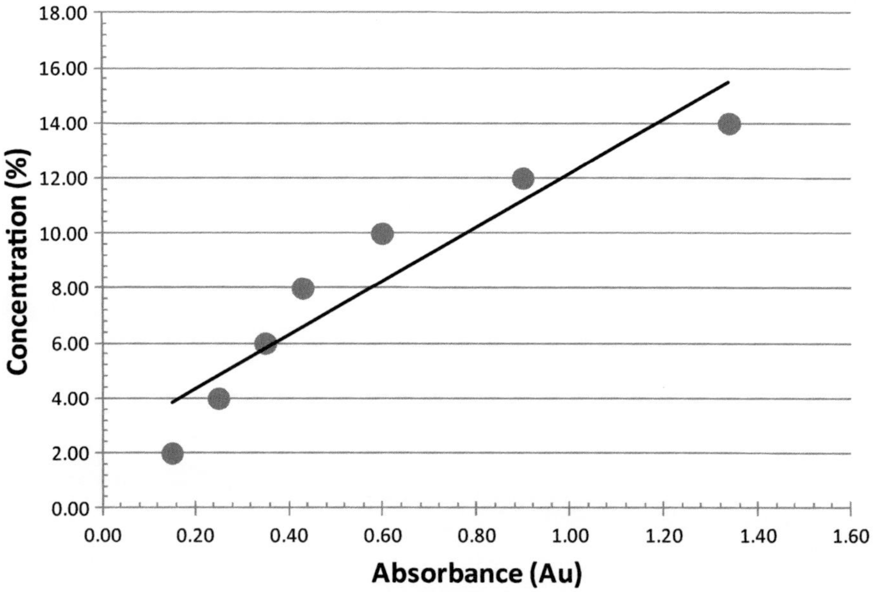

Fig. 6.13 Absorbance versus actual (reference) concentration data plot for a linear regression fit to the curved or non-linear data from Table 6.3. An unacceptable correlation is shown with relatively poor fitness for the absorbance values versus reference concentration values.

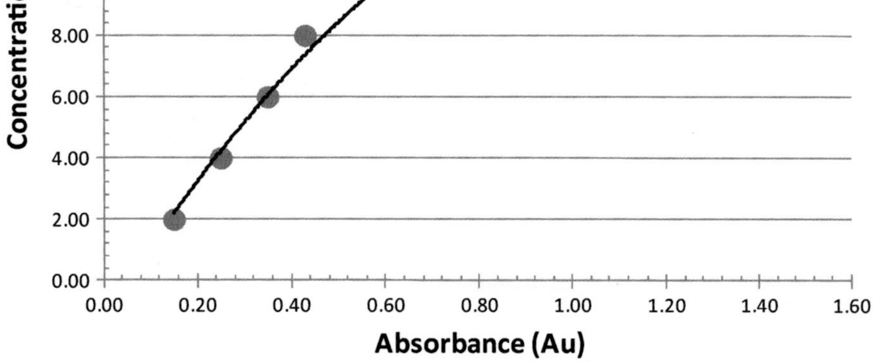

Fig. 6.14 Absorbance versus actual (reference) concentration data plot for a polynomial regression fit to the curved or non-linear data from Table 6.3. A high correlation is shown with relatively good fitness for the absorbance values versus reference concentration values.

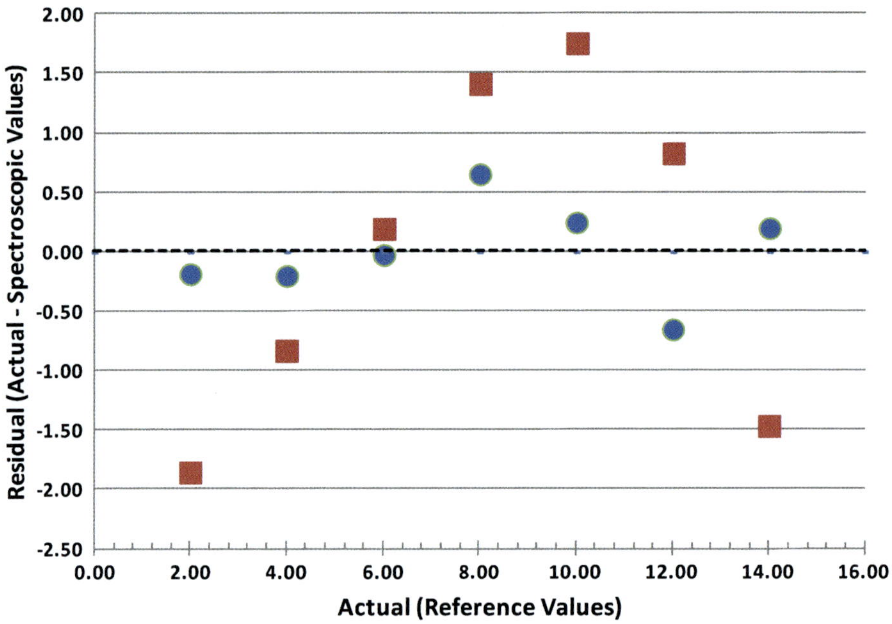

Fig. 6.15 Actual versus residual plot for linear and polynomial regression fitted equations for curved, non-linear data from Table 6.3. The linear fit has unacceptably high residuals, while the polynomial regression shows a strong positive correlation with excellent (precision), no slope or bias differences are shown; no curvature to data for residual values versus reference values is demonstrated. (Linear fit residuals are brown squares, polynomial fit residuals are blue circles.)

A comparison of the linear and polynomial regressions are illustrated as a residual plot in Figure 6.15.

6.3 Data Preprocessing: Scaling, Smoothing, and Normalization

Multiple methods are used for preparation of spectroscopic data either for visual evaluation or application to algorithms for qualitative or quantitative analysis. These methods are illustrated and described throughout these volumes. The reader is referred to the master index for additional details on each topic.

6.3.1 *Scaling Methods for Spectroscopic Data*

6.3.1.1 *Mean Centering*

The mean spectrum for a set of spectra is subtracted from all the spectra in the set. This technique is used to remove the larger (or bulk) spectral features in order to view or calibrate the remaining spectral information. This remaining information contains the detailed chemical features of the spectra. This approach is useful for most spectral libraries when looking for fine structural differences, or when applying multivariate calibration to a set of very similar samples.

6.3.1.2 *Autoscaling*

The mean spectrum of a set of spectra is subtracted from all the spectra (mean centering) with an additional step of dividing all the spectra in the calibration set by the standard deviation of absorbance values at each wavelength (datapoint). This technique is used to enhance the basic chemical information in spectra for manual viewing or for use in multivariate analysis, and is somewhat useful for reducing the scattering effects of solid or particulate samples.

6.3.2 *Smoothing Algorithms*

The process of smoothing spectroscopic data reduces or eliminates the white noise within spectra due to both random and structured variation in signals. Smoothing allows enhanced manual viewing and multivariate processing by eliminating spectral variation due to random measurement noise. Details of Savitzky-Golay smoothing and other smoothing methods in these volumes are found by referring to the master index. In the following section there is a brief description of the most common applications of smoothing within spectroscopy. Each of these topics may be also located within the master index for the volume set.

6.3.2.1 *Boxcar Smoothing*

Boxcar smoothing is an algorithm used to improve the signal-to-noise ratio in spectral data by averaging successive data points to remove spurious random variation or white noise effects. The use of this smoothing may also broaden bands and remove some fine structure or eliminate some poorly resolved small spectral bands within the data. Over smoothing will remove both important spectral features and create an x-axis (i.e., wavelength) shift in spectral data. Note that basic smoothing is a way to reduce the apparent resolution or bandwidth/linewidth of the spectra. For example, high resolution spectral data may be made to mimic lower resolution data by the application of smoothing.

6.3.2.2 *Fourier-Domain (FFT) Smoothing*

Fourier-domain smoothing involves a Fourier transformation of the signal, the application of a filter function (with a set filter factor), and then performing a reverse Fourier transform to the data. A triangular filter function is generally applied.

The FFT smoothing is able to remove high frequency random noise from the signal with a minimum effect on real spectral features. If overfiltering is applied additional spectral artifacts may appear, which have no chemical meaning and may be misinterpreted as real or actual chemical information. There is significant information on this technique found within these volumes, refer to the master index.

6.3.3 *Normalization*

6.3.3.1 *Normalization by Means of a Pathlength Correction*

For normalization, individual pathlength data is used as a scalar multiplier term for correction of an individual spectrum or group/set of spectra of varying pathlengths or varying effective pathlengths.

6.3.3.2 *Normalization by Area*

For spectral normalization by area, all band areas are set to a single integrated area value for the purpose of visual signal comparison and/or correlation techniques for all absorption bands present within the spectra. It is most useful for qualitative or identification searching for the presence of various chemical components within a mixture spectrum. It is also useful for spectral qualitative analysis when directly comparing spectral shapes of mixtures of materials. Refer to the master index for illustrations of this technique within these volumes.

6.3.3.3 *Normalization by Means of a Reference Band*

For applying normalization by means of a reference band, a set of spectra used from a multivariate calibration are ratioed to a peak height or are ratioed by a factor for the peak area of an isolated reference band. The reference band must be of consistent height or area and when ratioed will correct the spectral signal for anomalies in the measurement conditions. The ratioed spectra will be corrected for use in quantitative analysis in cases where the spectral baseline is poorly resolved. Illustrations of this are included in these volumes (see master index).

6.3.3.4 *Normalization Using the Kubelka-Munk Transform*

The Kubelka-Munk transform (KMT) is applied as a linearization function to signals with scattering and absorptive characteristics as often encountered in diffuse reflectance. This relationship is defined as:

$$\text{Spectral Signal (KMT)} = \frac{k}{s} = \frac{(1 - R_{\lambda_i})^2}{2R_{\lambda_i}} \quad (6.6)$$

where k is the absorptive coefficient; s is the scattering coefficient; R is the reflectance signal from the sample; and λ_i is the wavelength where the signal is measured as Reflectance with respect to wavelength (i.e., the reflectance spectrum). Note that the standard absorbance transform ($A = \log_{10}(1/R)$) has been found to be as effective as other transforms, including Kubelka-Munk, for linearization of spectroscopic data for absorbance versus concentration when measuring samples using diffuse reflection.

6.4 Derivative and Smoothing Applications in Spectroscopy

6.4.1 *First and Second Derivative Nomenclature*

Note this discussion of derivatives for spectroscopic measurements begins with the common linearization of reflectance/reflection (R) or transmittance/transmission data (T) to absorbance (A) given by:

$$A = -\log_{10} R_l \quad \text{and} \quad A = -\log_{10} T_l \tag{6.7}$$

where A is absorbance expressed as reflectance or transmittance when R or $T = I/I_0$. This is identical to the forms of $A = \log_{10}(1/R)$ and $A = \log_{10}(1/T)$, respectively.

The absorbance ratio derivative method has been used for near infrared[1] and other spectroscopic methods and is given as the ratio of absorbance at one wavelength divided by the absorbance at a second wavelength. This ratio method is also common to infrared spectroscopists; as its application results in a term which compensates to some extent for baseline offset. The ratio of the straight transmittance or reflectance data is also useful for normalizing or reducing baseline offset in spectroscopic measurements. For absorbance data, a typical absorbance ratio is given as:

$$\frac{A_{\lambda_1}}{A_{\lambda_2}} \tag{6.8}$$

Standard derivative nomenclature for first and second order derivatives is as Equations 6.9 and 6.10, respectively.

$$\text{First derivative} = f'(x) = \frac{dy}{dx} \tag{6.9}$$

$$\text{Second derivative} = f''(x) = \frac{d^2y}{dx^2} \tag{6.10}$$

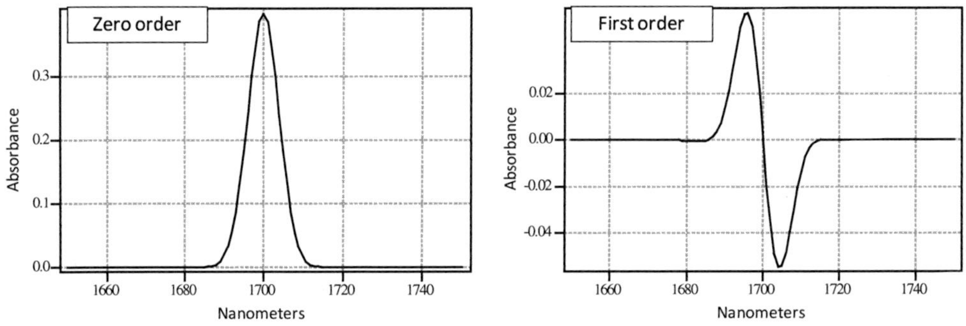

Fig. 6.16 Spectral shapes as zero order Gaussian band (left) and first derivative of Gaussian band (right).

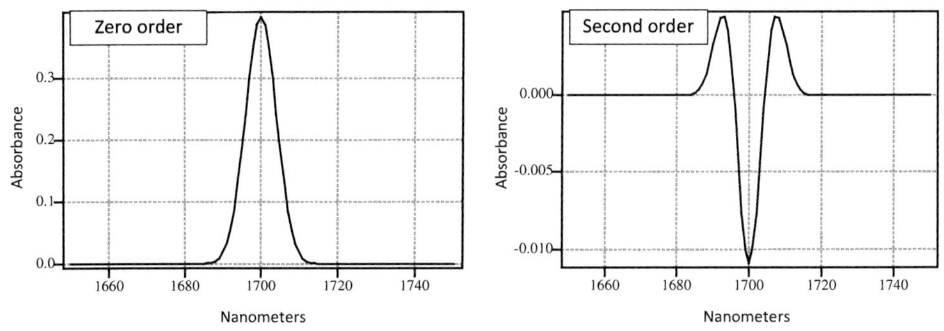

Fig. 6.17 Spectral shapes as zero order Gaussian band (left) and second derivative of band (right).

A first-derivative spectral plot as compared to an absorbance band (or single Gaussian band) is shown in Figure 6.16. The second derivative of the same Gaussian band is shown in Figure 6.17.

The use of the moving-averaged-segment convolution (MASC) method for computing derivatives in spectroscopy, brings about the expression below describing a first derivative term as:

$$A_{\lambda_2} - A_{\lambda_1} = A_{\lambda_{1+\Delta\lambda}} - A_{\lambda_{1-\Delta\lambda}} \tag{6.11}$$

where the derivative is given as the difference in absorbance values at 2 different wavelengths; the position of each wavelength being determined as the gap ($\Delta\lambda$) distance as + or − a specific wavelength or datapoint gap from a center wavelength (λ_1).

The MASC form of second derivative data processing is shown below.[2,3] In this case, the second derivative term is defined by the sum of absorbances at two wavelengths ($\lambda_{1+\Delta\lambda}$, and $\lambda_{1-\Delta\lambda}$), minus 2 times the absorbance at a center wavelength (λ_1). In this case, the second derivative gap size is designated as $\Delta\lambda$ (delta wavelength). The second derivative preprocessing step is quite effective in removing slope and offset variations in spectral measurement baselines. It also "assists" the calibration mathematics in defining spectral regions where small response changes can be useful in predictive calibration modeling. Without the use of derivatives, these spectral regions would not be beneficial for use in predictive calibration modeling as Equation 6.12.

$$A_{\lambda_1} + A_{\lambda_3} - 2A_{\lambda_2} = A_{\lambda_{1-\Delta\lambda}} + A_{\lambda_{1+\Delta\lambda}} - 2A_{\lambda_1} \tag{6.12}$$

Other derivative forms and formats are used to express computations of derivative spectra. These are a general form of the derivatives used in derivative processing software for spectrometers. These include the first derivative spectrum as a central difference method (Equation 6.13).

$$A'_{\lambda} = \frac{A_{\lambda+\Delta} - A_{\lambda}}{\lambda_1 - \lambda_2} = \frac{A_{\lambda+\Delta} - A_{\lambda}}{\Delta\lambda} \tag{6.13}$$

Where Δ is the number of wavelength spaces between A_λ and $A_{\lambda+\Delta}$.

A central difference method for computing the first derivative spectrum is used to mitigate for the typical shifts resulting from computing the derivative of spectral data using other equation forms. For this nomenclature the $\lambda + \Delta$ represents a center wavelength plus a wavelength gap, represented by (Δ). Therefore, λ-Δ represents the central wavelength minus the designated wavelength gap as Equation 6.14.

$$A'_{\lambda} = \frac{A_{\lambda+\Delta} - A_{\lambda-\Delta}}{2(\Delta\lambda)} \tag{6.14}$$

The method for computing the second derivative spectrum computation (Figure 6.17) is given as Equation 6.15.

$$A''_{\lambda} = \frac{(A_{\lambda+\Delta} - A_{\lambda}) - (A_{\lambda} - A_{\lambda-\Delta})}{(\Delta\lambda)^2} = \frac{(A_{\lambda+\Delta} - 2A_{\lambda} + A_{\lambda-\Delta})}{(\Delta\lambda)^2} \tag{6.15}$$

Where Δ is the number of wavelength spaces (gap) between A_λ and $A_{\lambda+\Delta}$ and $A_{\lambda-\Delta}$.

6.5 Simple Smoothing Function

6.5.1 *Savitzky-Golay Derivative Function of Instrument Software*

Derivative spectra allow enhanced visual identification of small spectral features. Applying first- through fourth-order derivatives allow the analyst to observe and study small spectral features which may not be apparent using direct absorbance or transmittance spectral data. One of the best known procedures for producing derivative spectra is the Savitzky-Golay convolution method (Analytical Chemistry journal, Volume 36, 1627, 1964).[4,5] This is the method used with many scanning spectrophotometers as the appropriate method for computing the derivative spectra with smoothing.

When using the Savitzky-Golay convolution method with continuous spectral data, the user may select an additional set of functions to include first- through fourth- derivative, and a sensitivity setting. The sensitivity function represents a derivative sensitivity, or data interval of 1 nm through 10 nm; each sensitivity number representing 1 nm. This technique is described in significant detail within other volumes, please refer to the master index for location of these sections.

6.6 Ultraviolet Instrument Optical Quality Performance Tests

The following series of tests is used to qualify or validate ultraviolet instrument performance and to determine which issues are problematic due to deficiencies in instrument alignment or calibration functions. These tests are related to alikeness in measurement performance between instruments and to overall precision and accuracy, as repeatability and reproducibility, respectively. Note that the terms optical density (O.D.) and absorbance units (A, Au, or AU) are synonyms. (Note that "Optical density" is found in historical documents, and is still used in physics, and biomedical and optical engineering, but not often in analytical chemistry.) Standard methods and practices for instrument calibration include multiple standards such as ISO 17025, ASTM E275, ASTM E169, ASTM E925, and ASTM E958.[6–10]

6.6.1 *Wavelength Accuracy Test*

Verify the wavelength accuracy of the spectrophotometer using suitable reference standards. The results must be consistent with the instrument manufacturing specifications, and for the application in use. For ultraviolet spectrometers, reference standards used include glass materials with embedded rare earth oxide materials, and liquids under standard measurement conditions. For both ultraviolet and visible wavelength accuracy and precision tests, materials used include those shown in Tables 6.4 and 6.5, the nominal (measurand) wavelengths given for wavelength calibration emission lamps, and solid materials used for testing alignment, respectively. Note that the percent uncertainty is computed as follows.

$$\text{Uncertainty}\ (\%) = [(\text{Measured Wavelength - NIST})/\text{NIST}] \times 100\% \qquad (6.16)$$

Table 6.4 Calibration emission lamp standards used for UV-Vis wavelength testing.

Lamp type	Nominal wavelengths (in nm)	Specified uncertainty (± % from NIST)	Allowed deviation (± nm)
Deuterium lamp	486.0	1.0%	1 nm
	656.1	1.0%	3 nm
Mercury vapor lamp	253.7	1.0%	1 nm
	302.25	1.0%	1 nm
	313.16	1.0%	1 nm
	334.15	1.0%	1 nm
	365.48	1.0%	3 nm

Table 6.5 Standards used for UV wavelength testing.

Material	Nominal wavelengths in nm	Specified uncertainty (± % from NIST)	Allowed deviation (± nm)
NIST SRM 2035a	334.6	0.2 %	0.7 nm
	345.4	0.2%	0.7 nm
	360.8	0.2%	0.7 nm
	374.5	0.2%	0.7 nm
	386.1	0.2%	0.7 nm
Holmium oxide quartz filter	279.2	1.0%	1 nm
	360.9	1.0%	3 nm
Holmium oxide (4%) in perchlorate ($HClO_4$) solution (10%)	241.15	1.0%	1 nm
	287.15	1.0%	1 nm
	361.5	1.0%	3 nm
	249.9	1.0%	—
	278.2	1.0%	—
	333.5	1.0%	—
	345.6	1.0%	—
	361.4	1.0%	—
	385.4	1.0%	—
Didymium Oxide in perchloric acid	298.0	1.0%	—
	329.0	1.0%	—
	354.0	1.0%	—
Samarium Oxide in perchloric acid	235.2	1.0%	—
	266.7	1.0%	—
	279.1	1.0%	—
	280.2	1.0%	—
	305.3	1.0%	—
	317.6	1.0%	—
	331.7	1.0%	—
	344.6	1.0%	—
	401.5	1.0%	—

and further U (Uncertainty) is given as:

$$\text{Uncertainty (nm)} = (\text{Measured Wavelength - NIST}). \tag{6.17}$$

For a more detailed NIST uncertainty value, the following equation is used. NIST uncertainty is a true measure of uncertainty unlike the typical instrument manufacturer report as either precision or accuracy. For example, NIST uncertainty has been calculated from Equation 6.18.[11]

$$U = \sqrt{A^2 + B^2 + C^2} \tag{6.18}$$

Where A = twice the largest standard deviation of measurement of multiple emission lines (measurands) versus the spectrometer measured values over a period of 31 days (or some designated period); B = twice the standard deviation of the uncertainty in the peak selection method used; and C is the maximum variation in the standard due to temperature changes over a specified range and conditions. (Note: the number 2 (twice) is referred to as the k value also known as the *coverage factor*, see reference 12.)

A list of NIST standard materials for calibrating UV-Vis spectrophotometers is included in the following Table 6.6.[13]

For the wavelength accuracy test, take repeat measurements of the reference sample (i.e., emission lamp, filter, or liquid sample) by placing it in the sample beam and not mechanically moving the sample over a normal measurement cycle for the instrument. For example, a 10-30 second measurement cycle may be typical. Then calculate the first derivative of each of the replicate spectra — compute the inflection or zero-crossing positions for the center band at the peak absorbance near the reference wavelength position (λ_{ref}) specified (see above tables of nominal

Table 6.6 NIST standard reference materials (SRMs) for calibrating UV-Vis spectrometers.

SRM No.	Description	Unit size	Wavelength range (in nm)
931g	Liquid Absorbance Filters, UV-VIS	set (12)	302 to 678
935a	Potassium Dichromate, UV Absorbance Standard	15 g	235 to 350
1935a	Potassium Dichromate Solution/UV Absorbance Standard	set (10)	235, 247, 313, 350
2031b	Metal-on-Fused-Silica Neutral Density Filters (250 nm-635 nm)	set (3)	250 to 635
2032	Discontinued by NIST	—	—
2034	Discontinued by NIST	—	—
2035a	Ultraviolet-Visible-Near-Infrared Wavelength/ Wavenumber Transmission Standard	each	334 to 805 and 971 to 1960

or measurand values for emission lamps and reference standards [references 13 and 14]). Next, calculate the standard deviation of difference of the wavelength positions for the zero crossings for scan-to-scan within (n) replicate samples, and the mean spectrum position ($\bar{\lambda}_i$) for the measured (λ_{ij}) versus reference (λ_{ref}) wavelength values. The standard deviation (s_i) is calculated as:

$$s_i = \sqrt{\frac{\sum_{j=1}^{n} (\lambda_{ij} - \lambda_{ref})^2}{n}} \quad (6.19)$$

The mean difference for wavelength accuracy is determined by:

$$\text{Wavelength Accuracy (as mean difference)} = \bar{\lambda}_i - \lambda_{ref} \quad (6.20)$$

where ($\bar{\lambda}_i$) is the average wavelength peak position for the scan-to-scan set; and (λ_{ref}) is the nominal (or measurand) reference wavelength position for the reference standard. The results are reported as Wavelength (λ) precision and accuracy as shown. This may be accomplished for multiple wavelengths depending upon the standard references measured.

	Nominal reference value (λ_{ref})	Precision (scan-to-scan standard deviation from reference, Eq. 6.19)	Accuracy (as mean difference from reference, Eq. 6.20)
Scan-to-Scan	—	—	—
Average	—	—	—

6.6.2 *Wavelength Repeatability Test*

Verify the wavelength repeatability of the spectrophotometer using suitable reference standard(s), such as those listed in the tables.

Take repeat measurements of the same emission lamp, filter, or liquid sample by placing it in the sample beam and not mechanically moving the sample over a normal measurement cycle for the instrument. For example, a 10–30 second measurement cycle may be typical. Then calculate the first derivative of each of the replicate spectra — compute the inflection or zero-crossing positions for the center band at the peak absorbance near the reference wavelength position (λ_{ref}) specified (see above tables of nominal or measurand values for emission lamps and standards). The standard deviation as wavelength repeatability is calculated as:

$$s_i = \sqrt{\frac{\sum_{j=1}^{n} (\lambda_{ij} - \bar{\lambda}_i)^2}{n-1}} \quad (6.21)$$

where (s_i) is the standard deviation for the scan-to-scan wavelength precision or repeatability for scan-to-scan measurements, and where (λ_{ij}) are individual wavelength shifts of the zero-crossover for sample i and scan-to-scan number j; $(\bar{\lambda}_i)$ is the average wavelength value for the scan-to-scan set; and n is the number of replicate measurements (pool all scan-to-scan data).

The mean spectrum wavelength position $(\bar{\lambda}_i)$ is calculated as:

$$\bar{\lambda}_i = \frac{\sum_i^n \lambda_{ij}}{n} \tag{6.22}$$

The results are reported as: Wavelength (nm) Repeatability

	Mean wavelength $(\bar{\lambda}_i)$, Eq. 6.22	Precision/Repeatability (s_i), Eq. 6.21
Scan-to-Scan	—	—

6.6.3 *Absorbance/Response Accuracy Test*

Verify the response accuracy of the spectrophotometer using a suitable standard, for example a set of pre-specified reference neutral density (ND) filters (such as NIST SRM 935a and SRM 2031, see reference 13, as well as Chapter 8 of this volume) with nominal (measurand) absorbance unit values. This reference standard must be provided with reference measurements of two or more separated wavelength positions.

For this test, repeat measurements of the same photometric standard are completed by placing it in the sample beam and not mechanically moving the sample over a normal measurement cycle for the measurement period. Then calculate the absorbance units (Au) for the entire spectrum and specifically at the two measured reference points of, for example, 235.0 nm, 257.0 nm, 313 nm, 350 nm (for 20 mg/L to 100 mg/L potassium dichromate), and SRM 935a. See index and Chapter 8 in this volume for more details regarding standard reference materials, their spectra and certified values.

The reference AU for each wavelength position (A_{235nm} and A_{350nm}) is measured for each scan for each sample (n) and for the mean spectrum $(\bar{A}_i)$ at each wavelength location over the measurement period. The statistics are calculated as Equation 6.23.

$$s_{235nm} = \sqrt{\frac{\sum_{j=1}^n (A_{ij} - A_{235nm})^2}{n}} \quad \text{and} \quad s_{350nm} = \sqrt{\frac{\sum_{j=1}^n (A_{ij} - A_{350nm})^2}{n}} \tag{6.23}$$

Where (s_{nm}) is the standard deviation (precision) for the scan-to-scan absorbance units at a specified wavelength (AU) accuracy for the set of measurements, and where (A_{ij}) are individual measurements of the absorbance units for sample i through scan-to-scan replicate measurement number j; (A_{nm}) are the reference values for the reference material at each of 235.0 nm and 350.0 nm; and n is the replicate measurement number. (Remember to pool all scan-to-scan data.)

The mean difference for photometric accuracy at each wavelength position is determined by:

$$\text{Photometric Accuracy (as mean difference)} = \bar{A}_{nm} - A_{nm} \tag{6.24}$$

where ($\bar{A}_{nm}$) is the average photometric value for the scan-to-scan set; and (A_{nm}) is the reference photometric value at each of 440.0 nm and 635.0 nm. The results are reported as Absorbance/Response (Photometric) Accuracy (in AU). These can be computed for multiple calibrated wavelength or wavenumber positions if desired. (Note: the blank items in the table below are completed for the analysis.)

	Nominal reference value A_{235nm}	Precision s_{235nm} Eq. 6.23	Accuracy (as mean difference) 235nm Eq. 6.24	Nominal reference value A_{350nm}	Accuracy as precision s_{350nm} Eq. 6.23	Accuracy (as mean difference) 350nm Eq. 6.24
Scan-to-Scan	—	—		—	—	
Average	—		—	—		—

6.6.4 *Absorbance/Response Repeatability Test*

Verify the response accuracy of the spectrophotometer using a suitable standard, for example a set of pre-specified reference neutral density (ND) filters (such as NIST SRM 930d, see reference 13) with nominal (measurand) absorbance unit values. This reference standard must be provided with reference measurements of two or more separated wavelength positions.

For this test, repeat measurements of the same photometric standard are completed by placing the standard in the sample beam and not mechanically moving the sample over a normal measurement cycle for the measurement period. Then calculate the absorbance units for the entire spectrum and specifically at the two measured reference points, for example 235.0 nm and 350.0 nm. The absorbance units are measured for each scan (i.e., scan-to-scan for each sample). Calculate the mean and standard deviation of the AU at the two wavelength positions for scan-to-scan (within replicate samples). This statistic is calculated for both 235.0 nm and

350.0 nm wavelengths as:

$$s_i = \sqrt{\frac{\sum_{j=1}^{n} (A_{ij} - \bar{A}_i)^2}{n - 1}} \tag{6.25}$$

where (s_i) is the standard deviation for the scan-to-scan absorbance units (AU) repeatability for the scan-to-scan measurements, and where (A_{ij}) are individual measurements of the absorbance for sample i and scan-to-scan number j; $(\bar{A}_i)$ are the mean measured values for the reference sample absorbance values at each of 235.0 nm and 350.0 nm; and n is the replicate number of spectra (pool all scan-to-scan data). The results are reported as follows. This can be repeated using a different calibrated set of wavelengths or wavelengths as needed.

Absorbance/Response (Photometric) Repeatability (in AU)

	Mean at 235 nm $\bar{A}_i$	Repeatability $s_{235\,nm}$ Eq. 6.25	Mean at 350 nm $\bar{A}_i$	Repeatability $s_{350\,nm}$ Eq. 6.25
Scan-to-Scan	—	—	—	—

6.6.5 *Stray Light Test*

The concept of stray light and the computations involved in computing stray light are described in the various volumes of this reference set; the reader is referred to the index and Chapter 8 of this volume for additional details. To measure the stray light in an instrument one is also referred to ASTM E387 test method.[15] This method uses two solutions to completely cut-off wavelengths of light from specific spectral regions, indicating that any light reaching the detector from those regions will be caused by stray light. The test solutions used are shown in Table 6.7.

The basic method to measure and compute stray light as percent transmittance of stray light is as follows. Place the cut-off filter (liquid) in the beam of the instrument and measure the scan in percent transmittance. Repeat the measurement by placing

Table 6.7 Solutions used to measure stray light and UV cut-off wavelengths.

Solution type	Cut-off Wavelength
Potassium Chloride, 12g/L Water	200 nm
Sodium Iodide (NaI), 10g/L Water	220 nm
Sodium Nitrite ($NaNO_2$), 50 g/L Water	340 nm and 370 nm
Acetone (pure HPLC grade)	250 nm to 320 nm

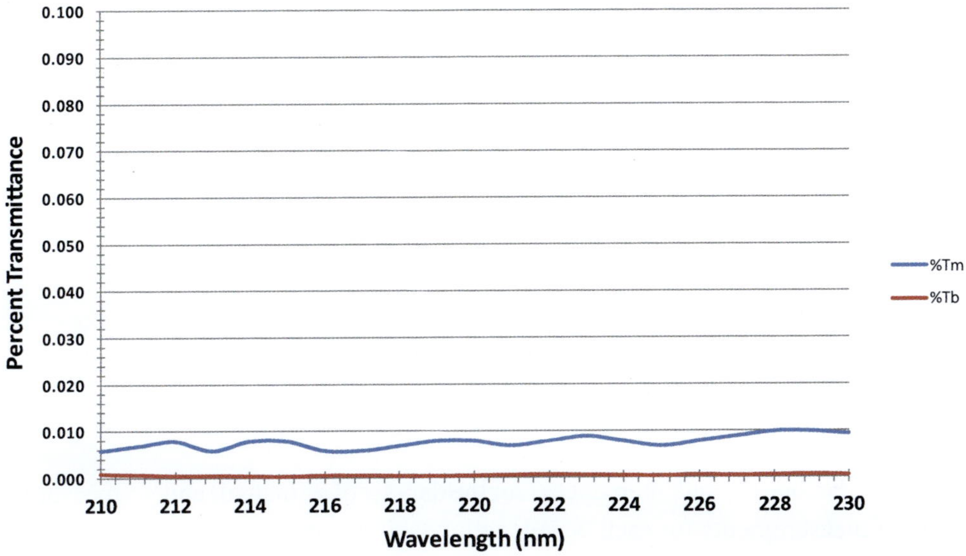

Fig. 6.18 Stray light measurement for cut-off filter near 220 nm. For the example shown, the stray light is less than 0.01 %T and is computed as 0.0080%T-0.0005%T = 0.0075%T at 220 nm.

a beam blocker in the instrument (this blocks all source lamp light from reaching the detector). The measurement should be made for a minimum of ±10 nm on either side of the cut-off wavelength for the filter or liquid used (e.g., 220 nm for Sodium Iodide solution), see Figure 6.18. Taking the percent transmittance data from both the filter and blocker measurements, use the equation as follows to compute the stray light (in %T):

$$\%T_S = \%T_m - \%T_b \tag{6.26}$$

where $\%T_S$ is the percent transmittance as stray light; $\%T_m$ is the measured percent transmittance using the cut-off filter; and $\%T_b$ is the percent measured transmittance using the beam blocker (should be 0%T).

6.6.6 *Photometric Linearity Test*

The photometric linearity of the spectrophotometer is tested using a set of neutral density reference filters. For this test, the measured response is plotted against the expected response. The slope of the line for reference (x) versus measured (y) data should be 1.00 ± 0.05 and the intercept 0.00 ± 0.05. Calculate the slope and intercept using the reference material measurements of absorbance such as 0.25, 0.50, 0.75, and 1.0 Au at 257, 416, or 630 nm for Starna Green; or by applying NIST SRM 936a for the establishment of a linearity test using Quinine Sulfate solution, respectively.

Quinine sulfate solutions should be linear over a concentration range of 0.25 to 1.0 mg/L in 0.105 mol/L of perchloric acid to a linear correlation (r) of 0.999. The useful wavelength range for Quinine Sulfate is 375 nm to 675 nm.

The results are reported as a graph of the measured linear response against the expected response at the two measured wavelengths; record full spectral data; and include the following table Linearity at Wavelength #1 ($\lambda 1$) and Wavelength #2 ($\lambda 2$). Other wavelengths may additionally be selected as needed.

Slope at $\lambda 1$ Eq. 6.27a	Intercept at $\lambda 1$ Eq. 6.27b	Slope at $\lambda 2$ Eq. 6.27a	Intercept at $\lambda 2$ Eq. 6.27b
—	—	—	—

Note the slope (b) and intercept (a) for the data set of x_i (actual) and y_i (measured) pairs of measurements for each wavelength position is given as:

$$b = \frac{\sum_i \{(x_i - \bar{x})(y_i - \bar{y})\}}{\sum_i (x_i - \bar{x})^2} \quad \text{and} \quad a = \bar{y} - b\bar{x} \qquad \text{(6.27a and 6.27b)}$$

6.6.7 *Photometric Noise Test*

Determine the photometric noise by using a reference neutral density filter. Repeat measurements of the same reference standard by placing it in the sample beam and not mechanically moving the sample over a typical measurement period. The photometric noise is computed as the standard deviation of the spectral response, and must be consistent with the instrument performance specifications for the application in use. This statistic is calculated for a single spectrum averaged over a standard measurement period as:

$$s_i = \sqrt{\frac{\sum_{j=1}^{n} (A_{ij} - \bar{A}_i)^2}{n - 1}} \qquad (6.28)$$

where (s_i) is the standard deviation (noise) for the *averaged* spectrum comprised of a number of scan-to-scan measurements for each wavelength, and where (A_{ij}) are individual absorbance measurements i for the averaged spectrum at wavelength j; ($\bar{A}_i$) is the average absorbance units value for the averaged spectrum; and n is the replicate number of spectra measured. This is completed for each wavelength or may be pooled for an entire spectrum. This can be reported at one or more wavelengths or over the entire measurement region.

The results are reported as: Photometric Noise

	Mean ($\bar{A}_i$)	Photometric Noise (S_i) Eq. 6.28
Averaged Spectrum	—	—

6.6.8 *Signal Averaging Test*

Determine the photometric noise by measuring a reference neutral density filter and *report results in transmittance*. Repeat measurements of the same reference standard by placing it in the sample beam and not mechanically moving the sample over the entire measurement period. This signal averaging test is to be completed using three methods:

6.6.8.1 *Random Noise Test*

This test excludes short-, medium-, and long-term drift; slope, and background curvature with time using measurements of alternating background and sample measurement spectra. This test simulates "dual beam" conditions, when using a single beam instrument, and excludes most of the impact from longer term periodic instrument drift, e.g., for n = 2, measure background, then measure sample, then background, then sample; reference each spectrum, then average the 2 referenced spectra — repeat this sequence for the appropriate number of co-added spectra: 1, 2, 4, 16, 64, 256, etc.; compute the background corrected spectra by referencing alternate (i.e., sandwiched sample and reference) spectra for averaged scans, then compute the standard deviation using Equation 6.28.

6.6.8.2 *Noise Test (Including Medium- or Short-Term Drift)*

Take background measurements of the same number of scans as sample measurement used for co-added result, for example: measure 1, 2, 4, 16, 64, 256, etc. as alternate background co-added set and then sample co-added set, e.g., for n=2, measure 2 scans as background followed in sequence by 2 scans of the reference sample, ratio these and calculate the standard deviation; for n = 4, measure 4 background and then 4 sample spectra, average these, and ratio as a single spectrum, continue this sequence and calculate the standard deviation using Equation 6.28. Note that a perfect, noise-free scan would be a flat straight horizontal line with no noise fluctuations.

6.6.8.3 *Noise Test (Including Long-Term Drift)*

Measure background at start of run and then measure samples in sequence using only the original background; this will be the only background scan used for calculations. Thus generate average spectra from r number of scans, across entire number of scans available, e.g., for n = 4: average scans 1-4, 5-8, etc.; for n = 16: average scans 1-16, 17-32, etc. Then calculate the standard deviation across the averaged spectra using Equation 6.28.

6.6.8.4 *Signal Averaging Test*

Signal Averaging Test — Obtain a series of replicate scan-to-scan spectra in transmittance or reflectance mode, and compute a subset of replicate scans and process them as described below. Do this for the following number of scans: 1, 4, 16, 64, 256, 1024, 4096, 16384, etc., up to the maximum measurement time of interest. Ratio each pair of sample and background spectra. (This is the same sample and reference measurement sequence that would be used for standard spectra data collection and is user selected.) With the data collected, calculate the noise level using Equation 6.28 at multiple wavelengths. Irrespective of the wavelengths selected for noise measurements, the noise level should be reduced by a factor of 2 for each successive ratioed spectrum; for example, if 1 scan gave a noise level of 1, 4 scans would give 1/2, 16 would give 1/4, 64 would give 1/8 and so on until this same level of noise reduction fails (i.e., failure of signal averaging test). The percent noise level for each successive ratioed spectrum should be a factor of 2 lower; for example, 1, 1/2, 1/4, 1/8, 1/16, 1/32, 1/64, 1/128, etc. (see results reporting table below).

Failure of Signal Averaging — Report the number of scans and the measurement time for each set of scan-to-scan data used in the particular ratioed spectrum and the noise level. Report a failure when the computed/measured noise level is a minimum of twice (2 times) that of the expected noise reduction. All spectrometers have a limit to their practical signal averaging capability, often set by residual interference fringing by optical components, by the apodization-determined feet of the water vapor interferences, by the electronic noise floor due to amplifier and detector performance, or mechanical spectrometer alignment or servo-motor errors.

The results are reported as Photometric noise signal averaging test at each wavelength, with a table reported for each of multiple wavelengths (preferably across the full spectral range).

Number of scans	Expected noise reduction factor	Measured photometric noise (s_i)	Measured noise reduction factor
1	1	—	—
4	1/2	—	—
16	1/4	—	—
64	1/8	—	—
256	1/16	—	—
1024	1/32	—	—
4096	1/64	—	—
16384	1/128	—	—
65536	1/256	—	—

6.6.8.5 *Resolution Test*

A basic resolution test is made by comparing the absorbance of a peak height to the absorbance at a peak trough and designating a threshold ratio as the measure of adequate resolution. For example, in ultraviolet-visible spectrometers there is a resolution test using the European Pharmacopoeia that defines adequate resolution for a spectrophotometer as determined using the absorbance ratio for a solution of 0.02% W/v toluene in hexane. The peak for this solution is measured at 269 nm and the trough at 266 nm. The ratio for passing this test is >1.5. In summary, the ratio of absorbance at 269 nm divided by the absorbance at 266 nm must exceed 1.5 for passing the resolution test (Reference 16). A basic form of resolution testing is shown in Figure 6.19.

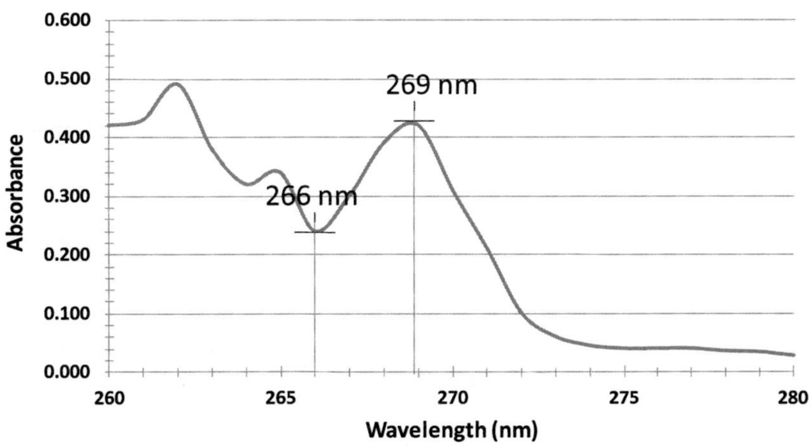

Fig. 6.19 Basic resolution test as a ratio of absorbance measurements for peak and trough of a spectral band.

6.6.9 *Summary Specifications for Instrument Performance Testing*

General specifications for comparing instruments should meet basic minimum pre-established criteria. These criteria are based on the actual use of the spectrophotometer and the accuracy required for measurements for a high signal to noise application. General requirements for each spectrometer should depend on the use, application, and overall performance requirements. The specifications here are used for illustrative purposes only. If extremely high performance and 'alikeness' is required then the repeatability or reproducibility of the instruments is most important. Here are some basic performance criteria that indicate a reasonable alikeness between two or more instruments. Specific calibrations and prediction performance criteria will vary and so then will the required comparative instrument performance metrics (Table 6.8).

6.7 Peak Selection Approaches (Peak Picking)[17–18]

There are multiple peak selection methods described in more detail within other volumes of this text series. Two of the most common methods are shown in this section. The reader is referred to the master index for additional information.

6.7.1 *Absolute Peak Maxima*

When using the absolute peak selection method find the maximum absorbance level, for absorbance spectra, or the minimum transmittance level, for transmittance

Table 6.8 Standard performance specifications for high quality ultraviolet spectrophotometer.

Specification or parameter	Expected specification
1. Wavelength accuracy	±0.20 nm
2. Wavelength repeatability	<0.05 nm
3. Absorbance/response accuracy	±0.004A at 1A
4. Absorbance/response repeatability	<0.001 AU (1 sigma)
5. Stray light test	<0.01 %T
6. Photometric linearity	Slope: 1.00±0.02; Intercept: ± 0.02 (Absolute)
7. Photometric noise	RMS <0.001 AU for 0.1%T ND filter
8. Signal Averaging Test	
8.1. Random Noise Test	Must pass method to 1024 scans
8.2 Noise Test (including medium- or short-term drift)	Must pass method to 500 scans
8.3 Noise Test (including long-term drift)	Must pass method to 480 scans
9. Resolution Test	A 269nm/A 266 nm> 1.5

spectra, and identify the peak position is the maximum point as illustrated in Figure 6.20.

6.7.2 *First Derivative Zero Cross-Over*

The maximum absorbance for any peak may be located using the first derivative cross-over point (Figure 6.21). A central difference method for computing the

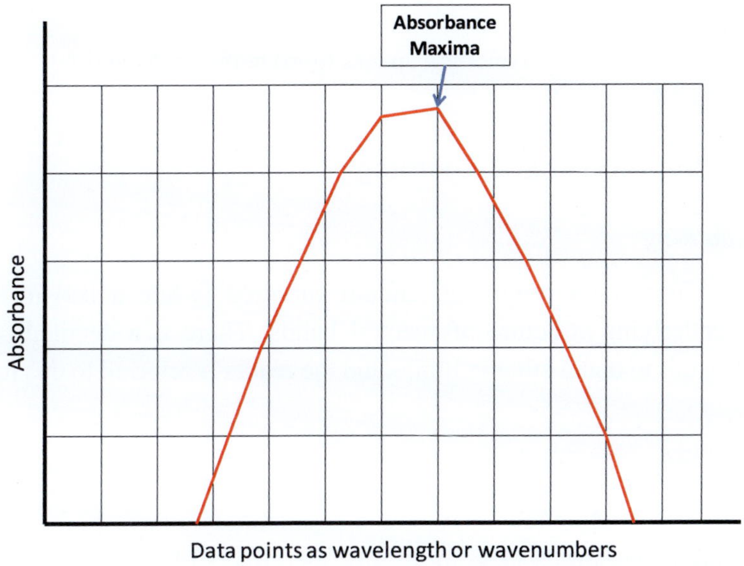

Fig. 6.20 Absorbance maxima as peak location.

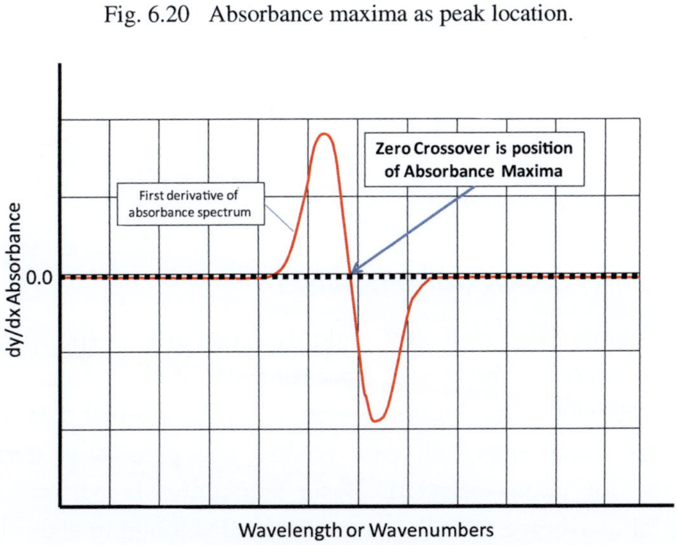

Fig. 6.21 First derivative cross-over point as the peak location

first derivative spectrum is used to mitigate for the typical shifts resulting from computing the derivative of spectral data using other derivative equation forms. For this nomenclature the $\lambda + \Delta$ represents a center wavelength plus a wavelength gap, represented by (Δ). Therefore λ-Δ represents the central wavelength minus the designated wavelength gap.

$$A'_\lambda = \frac{A_{\lambda+\Delta} - A_{\lambda-\Delta}}{2(\Delta\lambda)} \tag{6.29}$$

Where Δ is the number of wavelength spaces (gap) between A_λ and $A_{\lambda+\Delta}$ and $A_{\lambda-\Delta}$.

6.8 Data Preprocessing: Smoothing

6.8.1 *Smoothing*

Smoothing is useful for removing random variation (white noise) in spectra to reveal the underlying structure of spectral bands. There is a detailed section on smoothing in this text and other volumes and the reader is referred to this topic in the index.

6.8.2 *Derivatives*

Derivative preprocessing is used to remove the effects of offset between spectra, but not baseline slope or other effects due to particle size or energy scattering differences.[4,5,19] There is a special section on smoothing and derivatives to which the reader is referred (see index for smoothing or derivatives). Each successive derivative increases the noise by a factor of 10 and also adds structural artifacts into the spectrum based on the fine detailed structure of the spectrum undergoing derivative processing.

6.8.3 *Rectangular (Unweighted Smoothing)*

For basic smoothing (i.e., shift and multiply technique) a specified number of adjacent points from an absorbance spectrum are used to compute a point of the smoothed spectrum. The multiple points from the original data are combined to compute the single smoothed data points. The process is iterated for the specific number of points selected. Basic smoothing is performed using the rectangular sliding-average smooth (an unweighted smoothing algorithm). For this technique each of the smoothed data points is determined using the following

Equation 6.30; this example is for three smoothing points.

$$A_S = \frac{A_{\lambda-\Delta} + A_\lambda + A_{\lambda+\Delta}}{3} \tag{6.30}$$

Where A_S is the smoothed absorbance data point based on the absorbance values of three total data points. Note that $A_{\lambda-\Delta}$ (is the absorbance determined for the adjacent data point less 1 position); A_λ (is the absorbance for the original central data point to be replaced by the smoothed data computation (A_S)); and $A_{\lambda+\Delta}$ (is the absorbance data for the adjacent data point plus 1 position). These three original absorbance values are summed and divided by 3 (the number of data points to use in the smoothing calculation). Such a rectangular computational method may be derived for any number of smoothing points.

6.8.4 *Triangular (Weighted Smoothing)*

This smoothing technique is like the rectangular or unweighted smoothing method with the addition of weighting factors for giving more weight to the center data point and down weighting the influence of any adjacent points. This technique requires the divisor to be the sum of the weighting integers. So for a five point triangular smooth we use the Equation 6.31 as follows:

$$A_S = \frac{A_{\lambda-2\Delta} + 2A_{\lambda-\Delta} + 3A_\lambda + 2A_{\lambda+\Delta} + A_{\lambda+2\Delta}}{9} \tag{6.31}$$

where A_S is the smoothed absorbance data point based on the absorbance values of five total data points. Note that $A_{\lambda-\Delta}$ and $A_{\lambda-2\Delta}$ (are the absorbance values determined for the adjacent 2 data point less 1 and 2 positions, respectively); and A_λ (is the absorbance for the original central data point to be replaced by the smoothed data computation (A_S)). Note that $A_{\lambda+\Delta}$ and $A_{\lambda+2\Delta}$ (the absorbance data for the adjacent 2 data point plus 1 and 2 positions, respectively). The product and sum for these five original absorbance values are divided by 9 (i.e., the sum of the number of the weighting integers as 1+2+3+2+1=9). Such a rectangular computational method may be derived for any odd number of smoothing points.

6.9 X and Y Data Graphics (Plotting)

There are multiple conventions used for data plotting of spectroscopic results. These include standard methods for presenting reference and spectroscopic data comparisons.

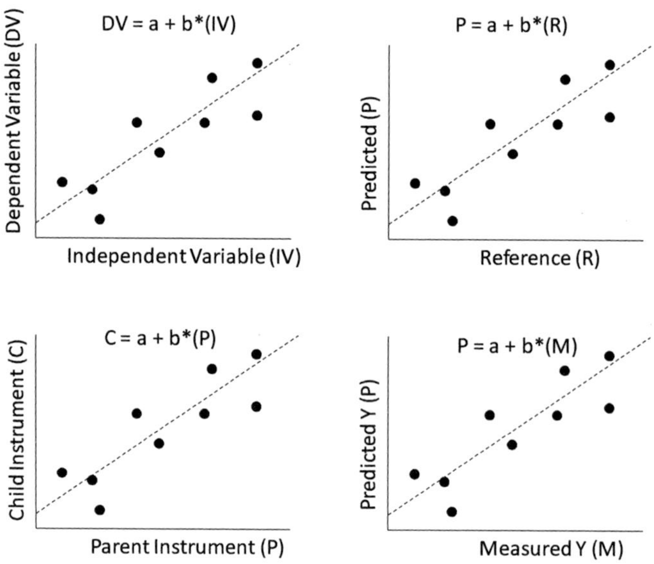

Fig. 6.22 A series of conventional X-Y plots as depicted in journal publications, textbooks and software for Independent Variable (X-axis) versus Dependent Variable (Y-axis), (upper left); Reference (Lab) Value versus Predicted (UV) Value (upper right); Parent (original calibration) Instrument versus Child (calibration transfer) Instrument (lower left); and Measured (Lab) Value versus Predicted (UV) Value (lower right).

6.9.1 *Conventional Data Plots*

Data plots of X versus Y have a variety of presentation formats based upon the application and context. The illustrated X and Y plots are used specifically and generically for spectroscopic data presentations (Figure 6.22).

6.10 ASTM Standard Methods for Ultraviolet Spectroscopy

Various practices and procedures are available from ASTM International that comprise consensus standards written by groups of experts and corrected or updated over time. These are listed below by their ASTM International designation numbers.[20–23]

ASTM E958-13, Standard Practice for Estimation of the Spectral Bandwidth of Ultraviolet-Visible Spectrophotometers.[20]

This practice provides methods and procedures for spectral bandwidth estimation of a spectrophotometer over the ultraviolet and visible region from 185 nm to 820 nm. The practice applies to conventional dispersive instruments as well as photodiode arrays.

ASTM E1866-97(2013), Standard Guide for Establishing Spectrophotometer Performance Tests.[21]

ASTM provides a guide for testing the performance characteristics for ultraviolet, visible, near-infrared, and mid-infrared spectrophotometers. This is a guide to supplement the existing Practices E275, E925, E932, E958, E1421, or E1683. This practice may be used to evaluate both laboratory and process (on-line) instrumentation.

ASTM E275-08(2013), Standard Practice for Describing and Measuring Performance of Ultraviolet and Visible Spectrophotometers.[22]

This is a set of test procedures used to describe the measurement performance of UV-Vis spectrophotometric instruments, most specifically for dispersive spectrophotometers used for transmittance measurements.

ASTM E169-04(2014), Standard Practices for General Techniques of Ultraviolet-Visible Quantitative Analysis

This practice is intended to provide general guidelines for applying UV-Vis spectroscopy for traditional quantitative analysis.

ASTM E925-09(2014), Standard Practice for Monitoring the Calibration of Ultraviolet-Visible Spectrophotometers whose Spectral Bandwidth does not Exceed 2 nm.[23]

This document provides a set of tests to evaluate spectral bandwidth for UV-Vis spectrophotometers capable of measuring over the range of 200 nm to 700 nm with an absorbance dynamic range specification from 0.0 to 2.0.

References

1. P. Williams and K. Norris (Eds.), *Near-Infrared Technology* (Amer. Assoc. Cereal Chemists, St. Paul, MN., 1987).
2. W.F. McClure and A.M.C. Davies, More on derivatives: Part 1: Segments, gaps and ghosts. *NIR News* **4**(6) (1994), 12.
3. W.F. McClure, More on derivatives. Part 2: Band shifting and noise. *NIR News* **5**(1) (1994), 14–16.
4. A. Savitzky and M.J.E. Golay, Smoothing and Differentiation of Data by Simplified Least Squares Procedures. *Anal. Chem.* **36**(8) (1964), 1627–1639.
5. A. Savitzky, "Data processing in analytical chemistry," *Anal. Chem.*, **33**(13) (1961), 25A–32A, 35A–42A.
6. ISO/IEC 17025:2005, General requirements for the competence of testing and calibration laboratories, International Organization for Standardization, ISO Central Secretariat, 1, ch. de la Voie-Creuse, CP 56, CH-1211 Geneva 20, Switzerland. Reviewed 2010.
7. ASTM E275-08(2013), Standard Practice for Describing and Measuring Performance of Ultraviolet and Visible Spectrophotometers, ASTM International, West Conshohocken, PA, 2013, www.astm.org.

8. ASTM E169-04(2009), Standard Practices for General Techniques of Ultraviolet-Visible Quantitative Analysis, ASTM International, West Conshohocken, PA, 2009, www.astm.org.
9. ASTM E925-09(2014), Standard Practice for Monitoring the Calibration of Ultraviolet-Visible Spectrophotometers whose Spectral Bandwidth does not Exceed 2 nm, ASTM International, West Conshohocken, PA, 2014, www.astm.org.
10. ASTM E958-13, Standard Practice for Estimation of the Spectral Bandwidth of Ultraviolet-Visible Spectrophotometers,ASTM International, West Conshohocken, PA, 2013, www.astm.org.
11. V. R. Weidner, P.Y. Barnes, and K.L. Eckerle, A Wavelength Standard for the Near Infrared Based on the Reflectance Of Rare-Earth Oxides, *J. Research NBS*, **91**(5) (1986), 243–253.
12. B.N. Taylor and C.E. Kuyatt, Guidelines for Evaluating and Expressing the Uncertainty of NIST Measurement Results, *National Institute of Standards and Technology (NIST), Technical Note 1297*, 20 pages (1994).
13. From https://www-s.nist.gov/srmors/ [Search UV].
14. Permission from Starna Ltd., Hainault, IG6 3UT, UK. www.starna.com
15. ASTM E387-04(2014), "Standard Test Method for Estimating Stray Radiant Power Ratio of Spectrophotometers by the Opaque Filter Method," ASTM International, West Conshohocken, PA, 2014, www.astm.org.
16. European Pharmacopoeia, Qualification of Equipment Annex 3: Qualification of UV-Visible spectrophotometers PA/PH/OMCL (07) 11 DEF CORR, 2007.
17. R.C. Fox, Computer searching of infrared spectra using peak location and intensity data, *Anal. Chem.* **48**(4) (1976), 717–721.
18. L. Brodsky, A. Moussaieff, N. Shahaf, A. Aharoni, and I. Rogachev, Evaluation of Peak Picking Quality in LC–MS Metabolomics Data, *Anal. Chem.* **82**(22) (2010), 9177–9187.
19. B.G. Osborne, T. Fearn, and P.H. Hindle, *Practical NIR Spectroscopy with Applications in Food and Beverage Analysis*. 2nd edn. (Longman Group, Burnt Mill, Harlow, Essex, England, UK, 1993).
20. ASTM E958-13, Standard Practice for Estimation of the Spectral Bandwidth of Ultraviolet-Visible Spectrophotometers, ASTM International, West Conshohocken, PA, 2013, www.astm.org.
21. ASTM E1866-97(2013), Standard Guide for Establishing Spectrophotometer Performance Tests, ASTM International, West Conshohocken, PA, 2013, www.astm.org.
22. ASTM E275-08(2013), Standard Practice for Describing and Measuring Performance of Ultraviolet and Visible Spectrophotometers, ASTM International, West Conshohocken, PA, 2013, www.astm.org.
23. ASTM E925-09(2014), Standard Practice for Monitoring the Calibration of Ultraviolet-Visible Spectrophotometers whose Spectral Bandwidth does not Exceed 2 nm, ASTM International, West Conshohocken, PA, 2014, www.astm.org.

Chapter 7

Applications and Spectral Examples for Ultraviolet Spectroscopy

Contents

This chapter describes the main uses of ultraviolet spectroscopy, including topics related to fluorescence spectroscopy. Other aspects of ultraviolet spectroscopy are also described. Please see the index or other chapters for specific topics related to calibrating ultraviolet spectrophotometers, measuring performance of spectrometers, and definitions of sample measurement geometry for ultraviolet measurements.

7.1 Ultraviolet Chromophores

A variety of applications are performed using ultraviolet spectroscopy. There are a number of specific chromophores measured using the UV region (up to 400 nm) and a variety of organic chemical functional groups as well; these are shown in Tables 7.1 and 7.2, and Figure 7.1. In addition, the designated cutoff wavelengths for various "spectrophotometric grade" solvents used for sample dilution in ultraviolet spectroscopy are given in Table 7.3.

7.2 Measuring the Sample for UV

The key aspects of measuring samples for ultraviolet measurement include both transmittance and reflectance measurement geometry. The ultraviolet spectral

Table 7.1 Ultraviolet chromophore band positions.

UV chromophore band positions	Start (nm)	End (nm)	Range (nm)
1,3 - Butadiene	213	220	7
1,3,5- Hexatriene	255	261	6
3-Butene-2-one (1st Band)	210	216	6
3-Butene-2-one (2nd Band)	317	323	6
Acetone (1st Band)	186	192	6
Acetone (2nd Band)	277	283	6
Acetone (in non-polar solvent)	276	281	5
Acetylenes (-C triple bond C-)	165	170	5
Alcohols (R-OH)	175	200	25
Aldehydes, saturated (R-C=O-H)	185	190	5
Aldehydes, unsaturated (C=C-C=O)	210	250	40
Alkenes (>C=C<)	175	185	10
Amide (in polar solvent)	211	217	6
Amides, primary (R-C=O-NH_2)	205	210	5
Amines, primary (1st Band) (R-NH_2)	185	190	5
Amines, primary (2nd Band) (R-NH_2)	200	220	20
Azo-group (R-N=N-R)	340	350	10
Carbolylic acids (in ethanol)	202	208	6
Carbonyl Group (R_2>C=O)	270	310	40
Carboxylic Acids (R-C=O-OH)	200	205	5
Esters (in polar solvent)	201	207	6
Esters (R-C=O-OR′)	200	205	5
Ethers (R-O-R)	180	185	5
Ethylene	175	180	5
Formaldehyde (in non-polar solvent)	290	296	6

See Figure 7.9 of this chapter for graphical representation of this table.

Table 7.2 Enzyme Related Ultraviolet Chromophore Band Positions in alphabetical order.[1]

Ultraviolet chromophore band positions	Start (nm)	End (nm)	Range (nm)
Enzymes: NADH and NADPH (Direct Absorption)	335	340	5
Enzymes: Lactate dehydrogenase (Direct Absorption)	345	350	5
Enzymes: NADH and NADPH (Direct Absorption)	335	340	5
Enzymes: Alcohol dehydrogenase (Direct Absorption)	335	340	5
Enzymes: Aldolase (Wavelength 1) (Direct Absorption)	235	240	5
Enzymes: Aldolase (Wavelength 2) (Direct Absorption)	335	340	5
Enzymes: D-Amino acid oxidase (Direct Absorption)	335	340	5
Enzymes: Catalase (Liberation of H_2O_2)	235	240	5
Enzymes: Cholinesterases (Color Reaction)	335	340	5
Enzymes: Alpha-Chymotrypsin (Hydrolysis Product)	275	280	5
Enzymes: Creatine kinase (Direct Absorption)	335	340	5
Enzymes: Deoxyribonuclease I (Depolymerization product)	255	260	5
Enzymes: Diamine oxidase (Direct Absorption)	335	340	5
Enzymes: Glucose-6-phosphate dehydrogenase (Direct Absorption)	335	340	5
Enzymes: Glucose phosphate isomerase (Direct Absorption)	335	340	5
Enzymes: Beta-Glucosidase (Direct Absorption)	335	340	5
Enzymes: Glutamate-oxaloacetate transaminase (Direct Absorption)	335	340	5
Enzymes: Glutamate pyruvate transaminase (Direct Absorption)	335	340	5
Enzymes: Gamma-Glutamyl transferase (p-nitroalinine method)	395	400	5
Enzymes: Hexokinase (Direct Absorption)	335	340	5
Enzymes: Alpha-Hydroxybutyrate dehydrogenase (Direct Absorption)	335	340	5
Enzymes: Isocitrate dehydrogenase (Direct Absorption)	335	340	5
Enzymes: Lactate dehydrogenase (Direct Absorption)	335	340	5
Enzymes: Acid phosphatase (Direct Absorption)	295	300	5
Enzymes: Pyruvate kinase (Direct Absorption)	335	340	5
Enzymes: Sorbitol dehydrogenase (Direct Absorption)	335	340	5
Enzymes: Trypsin (Direct Absorption)	275	280	5

measurement of both reference and test samples are made over the range of 190 nm to 400 nm. The spectrum must be accurate and repeatable for wavelength measurements (better than 0.2 nm accuracy); line width or bandpass (between 2–5 nm with high precision); and accurate in photometric intensity (better than 2–3% Reflection or Transmission as percent) as compared to first principles or standard reference samples of known wavelength and photometric qualities. These materials are described within Chapter 8. The typical measurement geometry types for ultraviolet spectrophotometers is given in the following Table 7.4 and illustrated in Figures 7.2–7.6.

The following figures illustrate the various sample measurement geometries for ultraviolet spectroscopy.

Ultraviolet Spectra Correlation Charts
From 52,632 cm^{-1} to 26,316 cm^{-1} (190 nm to 380 nm)

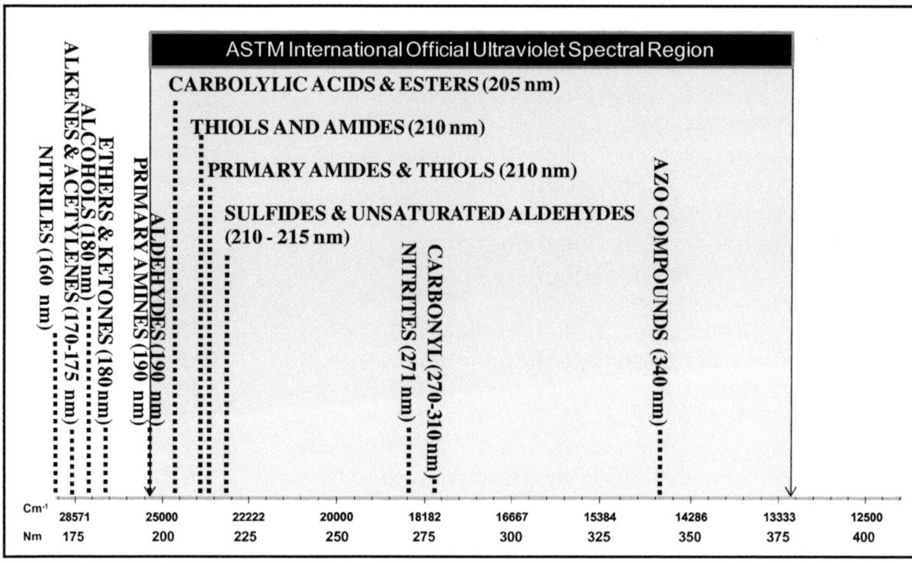

Fig. 7.1 Ultraviolet spectra correlation chart.

Table 7.3 Ultraviolet cutoff wavelengths for solvent materials.

UV cutoffs for solvents and optical materials	Start (nm)	End (nm)	Range (nm)
Acetic acid	260	265	5
Acetone	220	225	5
Acetonitrile	190	195	5
Barium Fluoride (BaF_2)	200	205	5
Benzene	280	285	5
l-Butanol	210	215	5
2-Butanol	260	265	5
Butyl acetate	254	259	5
Cesium Iodide (CsI)	250	255	5
Calcium Fluoride (CaF_2)	200	205	5
Carbon disulfide (CF_2)	380	385	5
Carbon tetrachloride (CCl_4)	265	270	5
1-Chlorobutane	220	225	5
Chloroform (stabilized with ethanol) ($CHCl_3$)	240	245	5
Cyclohexane	195	200	5
Diamond (C)	220	225	5
1,2-Dichloroethane	226	231	5
Dichloromethane	235	240	5
1, 2-Dimethoxyethane	240	245	5

(*Continued*)

Table 7.3 (*Continued*).

UV cutoffs for solvents and optical materials	Start (nm)	End (nm)	Range (nm)
N, N- Dimethylacetamide	268	273	5
N, N-Dimethylformamide	270	275	5
Dimethysulfoxide	265	270	5
1, 4-Dioxane	215	220	5
Diethyl ether	218	223	5
Ethanol	205	210	5
2-Ethoxyethanol	210	215	5
Ethyl acetate	255	260	5
Ethylene chloride	228	233	5
Glass, (BK-7) (SiO_2)	315	320	5
Glass, Flint (SF-10) (SiO_2)	380	385	5
Glass, Flint (SF-8) (SiO_2)	255	260	5
Glass, Optical Crown (SiO_2)	360	365	5
Glass, Pyrex®Borosilicate	360	365	5
Glass, Tempax®	360	365	5
Glycerol	207	212	5
Hexadecane	200	205	5
Hexane (n-Hexane)	201	206	5
Isooctane	195	200	5
Methacrylate	250	255	5
Methanol	205	210	5
2-Methoxyethanol	210	215	5
Methylcyclohexane	210	215	5
Methylethyl ketone	330	335	5
Methyl isobutyl ketone	335	340	5
2 -Methyl-1 -propanol	230	235	5
N-Methylpyrrolidone	285	290	5
Pentane	210	215	5
Pentyl acetate	212	217	5
Potassium Bromide (KBr)	250	255	5
1-Propanol	210	215	5
2-Propanol	210	215	5
Pyridine	330	335	5
Quartz, Natural crystalline (SiO_2)	240	245	5
Quartz, extremely low OH (SiO_2)	190	195	5
Sapphire (Al_2O_3)	150	155	5
Silica, Fused UV-Grade (SiO_2)	200	205	5
Silica, Standard-Grade (SiO_2)	230	235	5
Sodium Chloride (NaCl)	250	255	5
Suprasil 300®	190	195	5
Tetrachloroethylene (stabilized with thymol)	290	295	5
Tetrahydrofuran	220	225	5
Toluene	286	291	5
1, 1, 2-Trichloro-1, 2, 2-trifluoroethane	231	236	5
2,2,4-Trimethylpentane	215	220	5
o-Xylene	280	285	5
Water	190	195	5

Table 7.4 Typical measurement geometry for ultraviolet measurements.

Measurement geometry name	Angle of illumination*	Angle of detection*	Comments
Transmittance	0°	180°	Most Common
22.5°/22.5°	22.5°	22.5°	Common
Diffuse Reflection	0°	45°	Most Common
Specular Reflection	0°	<5°	Common

*Relative to Normal angle from the sample.

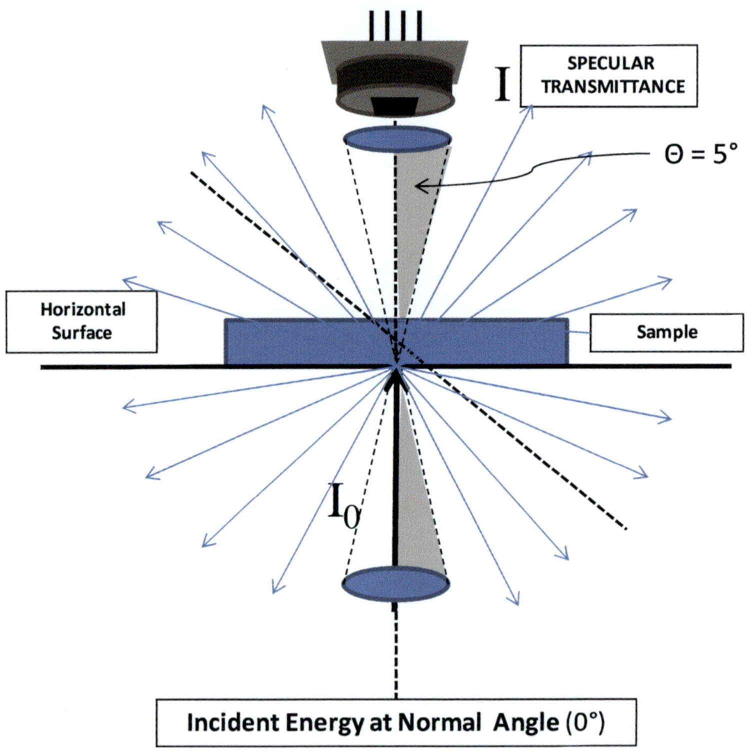

Fig. 7.2 0°/180° Transmittance/Transmission sample measurement geometry.

7.3 Reflectance Measurements Using Polarized Light

7.3.1 *Introduction*

Reflectance or transmittance measurements at varying angles, with or without linearly polarized light, are critical to the quality assurance measurements of

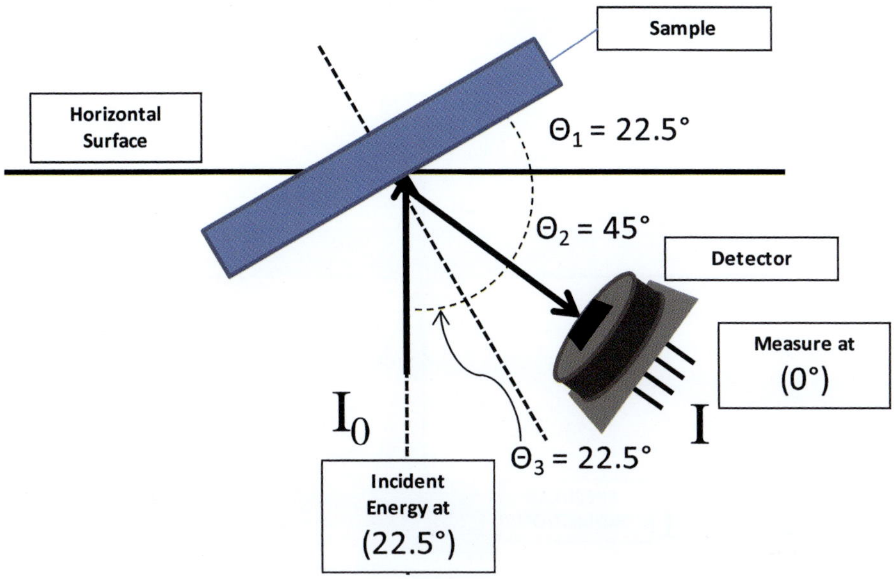

Fig. 7.3 A 22.5°/22.5° sample measurement geometry for specular reflection.

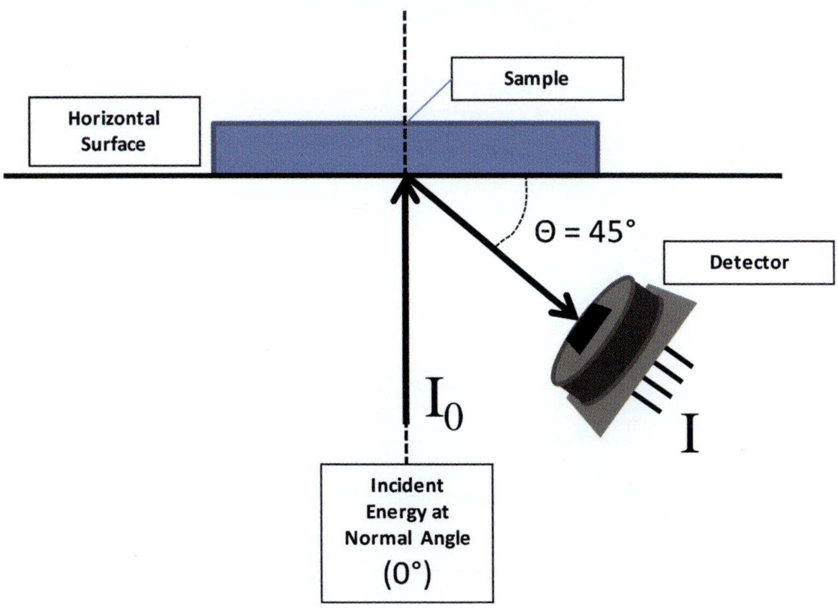

Fig. 7.4 The 0°/45° diffuse reflection sample measurement geometry.

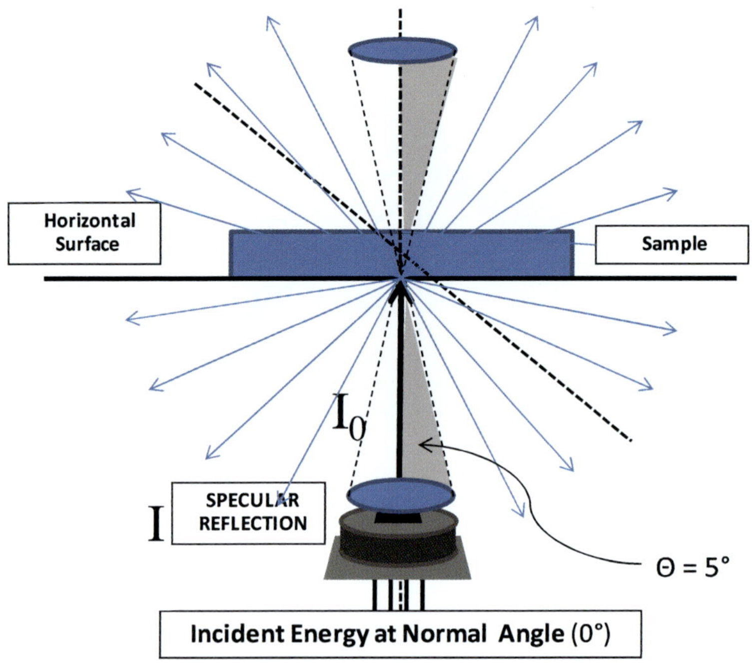

Fig. 7.5 The 0°/<5° specular reflection sample measurement geometry.

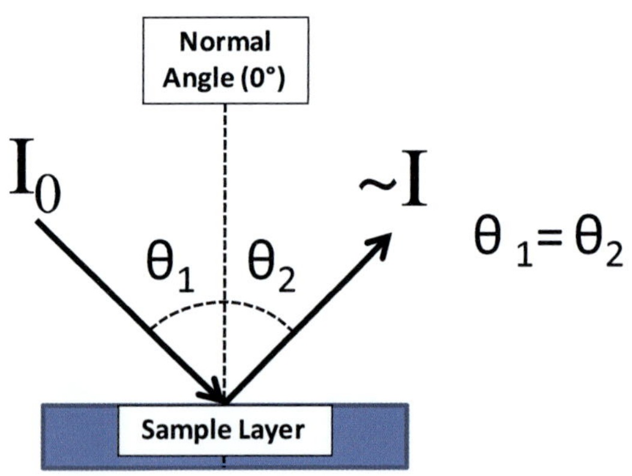

Fig. 7.6 In reflectance measurements the angle of incidence is equal to the angle of reflectance.

optical components, thin films, metal surfaces and semiconductor interfaces. For these materials, spectroscopic measurements of percent transmittance or percent reflectance are the most common uses of spectrophotometers for these applications. The angle of illuminance for both reflectance and transmittance is sometimes varied from near normal (approximately 0°) to 85° angles of illumination. For specular reflective surfaces the angle of incidence and reflection for directed light energy are equivalent as shown in Figure 7.6.

The smaller the angle of incidence of light interacting with a sample, the smaller is the degree of interaction with that surface. The electronic vector of the incident light determines the amount of interaction with a sample surface. When the electronic field vector of the surface is orthogonal (perpendicular) to the field direction vector of the incident energy, there is very little interaction with the surface. See Volume 2, Chapter 7 for more details on polarization measurements, as well as the master index.

7.4 Health Issues Associated with Exposure to Ultraviolet Radiation

7.4.1 *Threshold Limit Values (TLV) for Ultraviolet Exposure*

Ultraviolet radiation for health exposure concerns relates to the spectral region of 180 nm to 400 nm. The development of sunscreen products is for the purpose of mitigating direct exposure to the UV energy of sunlight. Note that the TLV for UV refers to conditions that all healthy humans may be repeatedly exposed to without acute adverse health effects. The health effects of greatest concern include erythema (sunburn), and photokeratitis (radiation exposure levels causing inflammation or rash of the skin). The following equations and discussion will describe the mathematics of TLV.

The computational equation for Effective irradiance, E_{ei} (or effective radiation exposure) is as follows:

$$E_{ei} = E_{si}\left[\frac{mW}{cm^2}\right] \cdot S_e[Unitless] \cdot t_{exp}[s] \tag{7.1}$$

where E_{si} is the spectral irradiance (as power or energy in mW/cm^2) at the center wavelength from the source; S_e is the tabled value for the relative spectral effectiveness at center wavelength (Table 7.5 and Reference 2); and t_{exp} is the allowed daily exposure time.

The daily TLV dose limit equation for daily exposure (in seconds) for ultraviolet (UV) radiation of a narrow band source is as follows. Note that E_{si} may be measured directly using a UV radiometer. The following Equation 7.2 assumes the daily exposure (t_{exp}) is limited to a maximum of 3.0 mJ/cm^2.

$$3.0\left[\frac{mJ}{cm^2}\right] \geq E_{si}\left[\frac{mW}{cm^2}\right] \cdot S_e[Unitless] \cdot t_{exp}[s] \tag{7.2}$$

Table 7.5 Ultraviolet Se and TLV values with respect to wavelength[2].

nm	S_e	TLV
180	0.012000	250
200	0.030000	100
220	0.120000	25
240	0.300000	10
260	0.650000	4.6
280	0.880000	3.4
300	0.300000	10
320	0.001000	2900
340	0.000280	11000
360	0.000130	23000
380	0.000064	47000
400	0.000030	100000

For example, if one has a source with a spectral irradiance (E_{si}) of 100 mW/cm^2 at 320 nm, the computation of the effective irradiance (E_{ei}) dose would be as Equations 7.3 and 7.4. Note: from Table 7.5, Ref. 2 for a 320 nm narrowband source, Spectral Sensitivity or Relative Spectral Effectiveness (S_e) is 0.0010; the unfiltered irradiance (E_{si}) is 100 mW/cm^2; and exposure times are 15 seconds and 30 seconds, respectively). So the computed E_{ei} values for these exposure times are:

$$100\left[\frac{mW}{cm^2}\right]\cdot 0.0010\cdot 15 = 0.15\left[\frac{mJ}{cm^2}\right] \tag{7.3}$$

$$100\left[\frac{mW}{cm^2}\right]\cdot 0.0010\cdot 30 = 3.0\left[\frac{mJ}{cm^2}\right] \tag{7.4}$$

So 30 seconds is the maximum safe daily exposure (i.e., ≤ 3.0 mJ/cm^2) for this UV source.

Furthermore, the maximum exposure time (in seconds) for this system may be determined using the following relationship as Equation 7.5.

$$t(max) = \frac{TLV\left[\frac{mJ}{cm^2}\right]}{E_{si}\left[\frac{mW}{cm^2}\right]} \tag{7.5}$$

Where t(max) is the maximum exposure time (in seconds) and TLV is the *Threshold Limit Values*. So for the UV source described above, one may compute the maximum

Table 7.6 Comparison of sources and energy levels[3–5].

Source	Irradiance (mW/cm^2)	Exposure time (s)	Type of source	Center λ (nm) of source	Bandwidth (nm) of source
Tanning Booth	0.09	138*	FL. Lamp	365	80
Dental Curing	1000+	10	LED	465	100
Solar	0.03	276*	Sun/Atmosphere	550	Broad/UV-IR

*The maximum initial recommended exposure time to prevent erythema as $0.5 \times MED$, using the UV irradiance for tanning lights and solar source.

safe daily exposure using a TLV of 2.9×10^3 mJ/cm^2 (Table 7.5, Ref. 2) for 320 nm.

$$t(max) = \frac{2.9 \times 10^3 \left[\frac{mJ}{cm^2}\right]}{100 \left[\frac{mW}{cm^2}\right]} = 29\, seconds \tag{7.6}$$

From this equation, the maximum safe daily exposure is 29 seconds.

7.4.2 *Irradiance Considerations Due to Distance from Target*

The effective irradiance should be determined at the target, not for the overall output of the source. This would take into consideration the source distance from the target, any lightguide materials, the addition of filters, lenses, or any other attenuating materials or angular radiance involved in the optical path.

Table 7.6 illustrates some common sources of ultraviolet radiation and typical maximum recommended safe daily exposure times.

7.5 Typical Applications of Ultraviolet Spectroscopy

There are many classic publications describing the application of ultraviolet spectroscopy for qualitative and quantitative analysis. The examples given are traditional and have existed for some time. A published document discusses multiple aspects of applying ultraviolet spectroscopy for measurements with theoretical references.[6]

A method is provided for transmittance measurement of monoethylene glycol (1,2-ethanediol; MEG) using ultraviolet spectroscopy over the spectral regions of 220 nm to 350 nm. The method provides a basis for determination of the purity of samples using a determination of extraneous ultraviolet absorbing compounds.[7]

The analysis or evaluation of organic paints and coatings for the degree of conjugated double or triple bonds and degree of aromaticity is determined using ultraviolet (UV) or visible (VIS) spectroscopy. Dyes, some colorants, and aromatic solvents may be evaluated using UV spectroscopy.[8]

Two- and three-dimensional fluorescence spectroscopic data have been recorded from sebum-rich latent fingermarks on quartz and from a standard white card. The fingermark residue will fluoresce over the emission range of 310 nm to 380 nm, with an excitation range between 260 nm to 300 nm. The emission data are used to image the ultraviolet photoluminescence of such latent fingermarks (fingerprints).[9]

A classic series of three papers present ultraviolet (UV) absorption data of drugs and toxic organic compounds systematically by grouping according to chemical structure. UV structural groups are spectrophotometrically characterized by an absorption profile from 200 nm to 340 nm. The spectra are reviewed and include the effect of pH change, decreased solvent polarity relative to water, and the overall intensity of specified absorption bands.[10–12]

Test methods are described using ultraviolet absorption spectroscopy of a variety of petroleum products. The test method describes the absorbance of liquids and solids for the spectral region of 220 nm to 400 nm.[13] A detailed test method describes a procedure for the examination and evaluation of National Formulary (NF) and United States Pharmacopeia (USP) grade white mineral oils.[14]

Direct analysis of oily water effluents was investigated as a method for the determination of petroleum hydrocarbons in the discharge from primary oil production facilities. A co-solvent, isopropanol (2-propanol), was chosen that produced a clear, homogeneous solution when added to oil-in-water effluent. The clear solution was then analyzed directly for aromatic hydrocarbon content by using ultraviolet spectroscopy. The concentration of oil in the sample was calculated from a Lambert-Beer plot at 226 nm. The standard used was a sample of oil that best represented the oil that had contacted the water, i.e., current production crude oil.[15]

Spectroscopy has been used to quantify the effects of ultraviolet light on a model polyester–urethane coating as it degraded in an accelerated exposure chamber. An explorative calculation of the effective dosage absorbed by the coatings was made and, depending on the quantum yield chosen to represent the damaging processes, was found to follow an accelerating trend with an exposure period that may permit estimation of a service lifetime by spectroscopy. The accelerating trend was due to an increase in the absorption "yellowing" tail as degradation proceeded. Quantification as an "Urbach" tail, and the underlying electronic transition by an analysis due to Tauc *et al.*, showed that the Urbach energy was consistent with values found elsewhere. It is possible that, for this urethane at least, any explanation of "yellowing" may not require the development of specifically yellow chromophores.[16]

Ultraviolet applications for conformational studies of proteins and amino acids have been described in detail.[17–18] Such literature discusses developments in the estimation of various conformations in a protein molecule from its circular dichroism (CD) spectrum in the ultraviolet region. Optical rotatory dispersion (ORD) and CD are two chiroptical phenomena, which differentiate two enantiomers. They are caused by different interactions of left- and right-circularly polarized light with chiral molecules. Chirality is a geometric property of molecules; the corresponding substances are therefore optically active. At present CD has replaced ORD for the conformational analysis of proteins because CD bands are characteristic of various secondary structures and can be directly observed. The very simplicity of the CD method and the short time required for the measurements are extremely attractive. It is often advantageous to compare the CD analysis with empirical predictions of secondary structure of proteins from sequences, if such are available. All current methods for CD analysis are developed to determine not only the helix and β-form but also the β-turn and unordered forms.[19]

7.6 Circular Dichroism (CD) using Ultraviolet Spectroscopy

Circular dichroism (CD) ultraviolet spectroscopy uses polarized ultraviolet light (as left-handed polarized light ratioed to right-handed polarized light) to measure asymmetry, or differences, in molecular structures of protein molecules. A zero CD measurement occurs for samples with regular or random structure, while positive and negative CD signals occur with highly structured or oriented molecules. A protein structural example is given in Figure 7.7 for random and helical (i.e., secondary and tertiary) protein structures.[20]

Note for protein structural analysis: (1) primary structures are represented by amino acid 'linear' sequences; (2) secondary structures include alpha-helices, beta-sheets, and beta-turns; (3) tertiary structures include helices and the folded helices; and (4) quaternary structures are represented as folding of tertiary structures and multi-subunit proteins.

7.6.1 *CD Analysis for Protein Structure*[21–23]

Ultraviolet absorptions in the 190 nm to 250 nm region may be used to measure protein secondary structure; in this wavelength region a folded peptide band structure may be observed. Random coil structural character has a CD positive signal near 212 nm, and the alpha-helix protein form has a positive signal near 190 nm. Beta-sheet protein structures have a positive band near 196 nm. Helical protein

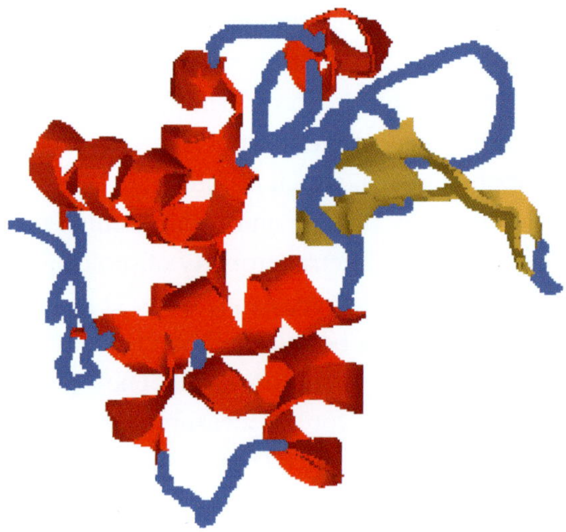

Fig. 7.7 Representative Protein Structure with α-helix (red), β-sheet (gold), and Random Coil (blue).

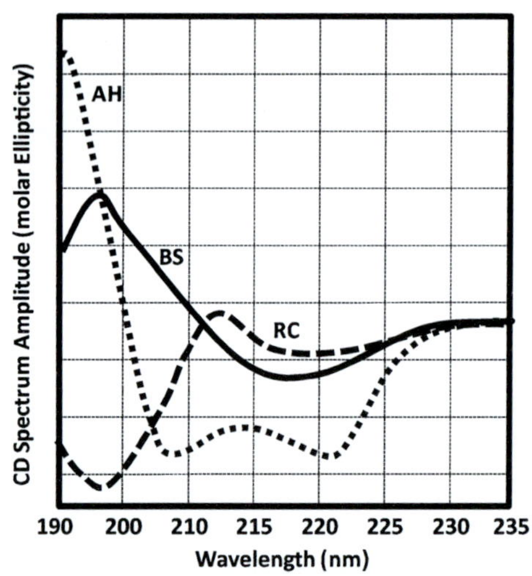

Fig. 7.8 Natural Protein CD Spectrum as Wavelength (nm) versus Amplitude (i.e., CD Molar Ellipticity in units of degrees $\cdot$ cm^2 $\cdot$ dmol^{-1}). Note that AH is Alpha Helix structure; BS is Beta Sheet structure; and RC is Random Coil structure. Figure is an illustration of Table 7.7.

structures include secondary structures (i.e., alpha-helices, beta-sheets, and beta-turns), and tertiary structure (i.e., helices and the folded helical structures). Note Figure 7.8. illustrating a CD analysis of natural protein. In this figure a protein CD spectrum amplitude (as Molar Ellipticity in units of degrees $\cdot$cm^2 $\cdot$ dmol^{-1}) versus wavelength is shown for the different protein conformational structures. Table 7.7

Table 7.7 UV-CD locations for various proteins and protein conformational features.

Protein or conformation	Far- or Near-UV	Wavelength (nm)	Comment(s)
α-helix	Far-UV (190–250 nm)	222 and 208 nm	Negative Band
α-helix	Far-UV (190–250 nm)	190 nm	Positive Band
β-sheet	Far-UV (190–250 nm)	218 nm	Negative Band
β-sheet	Far-UV (190–250 nm)	196 nm	Positive Band
Random Coil	Far-UV (190–250 nm)	195 nm	Negative Band
Random Coil	Far-UV (190–250 nm)	212 nm	Positive Band
Aromatic residues (AR)	Near-UV (250–350 nm)	250–350 nm	Positive Band
Tryptophan (AR)	Near-UV (250–350 nm)	280–300 nm	Positive Band
Tyrosine (AR)	Near-UV (250–350 nm)	270–290 nm	Positive Band
Phenylalanine (AR)	Near-UV (250–350 nm)	250–270 nm	Positive Band
Cystine disulfide (AR)	Near-UV (250–350 nm)	250–350 nm	Positive Band

shows detailed band positions for CD measurements of the specific protein structural configurations listed.

Other spectroscopic methods are used to determine detailed protein structures. For example circular dichroism for peptide amide bonds and aromatic residues uses far-UV and near-UV; secondary structures are often measured using infrared and Raman spectroscopy; NMR is used for detailed structural confirmation and is applied when sufficient sample (many milligrams) is available. When small sample milligram (mg) quantities are available, UV absorption or fluorescence measurements are often made.

UV-CD for protein structural determination is applied as follows. Secondary structure is determined using the peptide amide bond as the UV chromophore in the Far-UV (190–250 nm) region. Ultraviolet spectra include features for α-helix (negative bands at 222 and 208 nm, and positive band at 190 nm); β-sheet (negative band at 218 nm, positive band at 196 nm); and for random coil conformation (negative band at 195 nm, positive band at 212 nm) are used. Aromatic residues are determined using the existing residue chromophores in the Near-UV region (250–350 nm). These include tryptophan (280–300 nm), tyrosine (270–290 nm), phenylalanine (250–270 nm), and cystine disulfide (broad). See Table 7.7 for a summary of these protein group CD spectral locations.

Note that thermal stability (temperature change versus CD spectrum) and conformation differences (wavelength versus CD spectrum amplitude as Molar Ellipticity) between synthetic and natural purified proteins may be compared using ultraviolet CD. The CD spectrum is compared between the purified (known) natural protein and a synthetic protein. Significant differences in the comparison CD spectra indicate important protein conformational differences. These differences indicate

Table 7.8 Ultraviolet Functional Groups, λ_{Max} Band Positions, Intensities, and Transitions.

Functional Group(s) or *moiety*	Intensity	ε (absorptivity)	Wavelength (nm)	Transition
Amines, alcohols, ethers, thiols,	low to medium	100–10,000	<220 nm	$n \rightarrow \sigma*$
cyano (-C≡N:)	weak	<100	<220	$n \rightarrow \sigma*$
C=O, C=N, N=N, $-NO_2$, -COOR, -COOH, $-CONH_2$	very weak	10–100	250–360	$n \rightarrow \sigma*$
Aromatic system	medium	1000–10,000	>200	2 bands
Polynuclear Aromatics	medium	1000–10,000	>200	3 bands
α, β-unsaturated ketone	high	10,000–20,000	>210	—
Ketones, acids, esters, amides	low	100–5,000	>300	$n \rightarrow \pi*$ 2 bands
Ketones, acids, esters, amides	high	10,000	<250	$\pi \rightarrow \pi^*$ 2 bands
Quinoids (yellow colored)	high	10,000–30,000	350–500	broad 4–5 bands
Conjugated dienes (Woodward-Fieser Rules)	high	20,000–25,000	215–250	$\pi \rightarrow \pi^*$ 1–4 conjugated double bonds
Polyene systems (Fieser-Kuhn Rules) e.g., β-carotene	high	20,000–25,000	288, 449, 478, 518	—
Diene, Cis (Woodward-Fieser Rules)	high	1000–100,000	253	Parent diene
Diene, Cis (Woodward-Fieser Rules)	high	1000–100,000	283	Double bond extending conjugation
Diene, Cis (Woodward-Fieser Rules)	high	1000–100,000	258	Alkyl substituent or ring
Diene, Cis (Woodward-Fieser Rules)	high	1000–100,000	258	Exocyclic double bond
Diene, Cis (Woodward-Fieser Rules)	high	1000–100,000	253	$-OCOCH_3$ polar group
Diene, Cis (Woodward-Fieser Rules)	high	1000–100,000	259	-OR polar group
Diene, Cis (Woodward-Fieser Rules)	high	1000–100,000	258	$-Cl_2$, or - Br polar group
Diene, Cis (Woodward-Fieser Rules)	high	1000–100,000	313	$-NR_2$ polar group
Diene, Trans (Woodward-Fieser Rules)	high	1000–100,000	214	Parent diene
Diene, Trans (Woodward-Fieser Rules)	high	1000–100,000	244	Double bond extending conjugation
Diene, Trans (Woodward-Fieser Rules)	high	1000–100,000	219	Alkyl substituent or ring
Diene, Trans (Woodward-Fieser Rules)	high	1000–100,000	219	Exocyclic double bond
Diene, Trans (Woodward-Fieser Rules)	high	1000–100,000	214	$-OCOCH_3$ polar group
Diene, Trans (Woodward-Fieser Rules)	high	1000–100,000	220	-OR polar group
Diene, Trans (Woodward-Fieser Rules)	high	1000–100,000	219	$-Cl_2$, or - Br polar group
Diene, Trans (Woodward-Fieser Rules)	high	1000–100,000	274	$-NR_2$ polar group

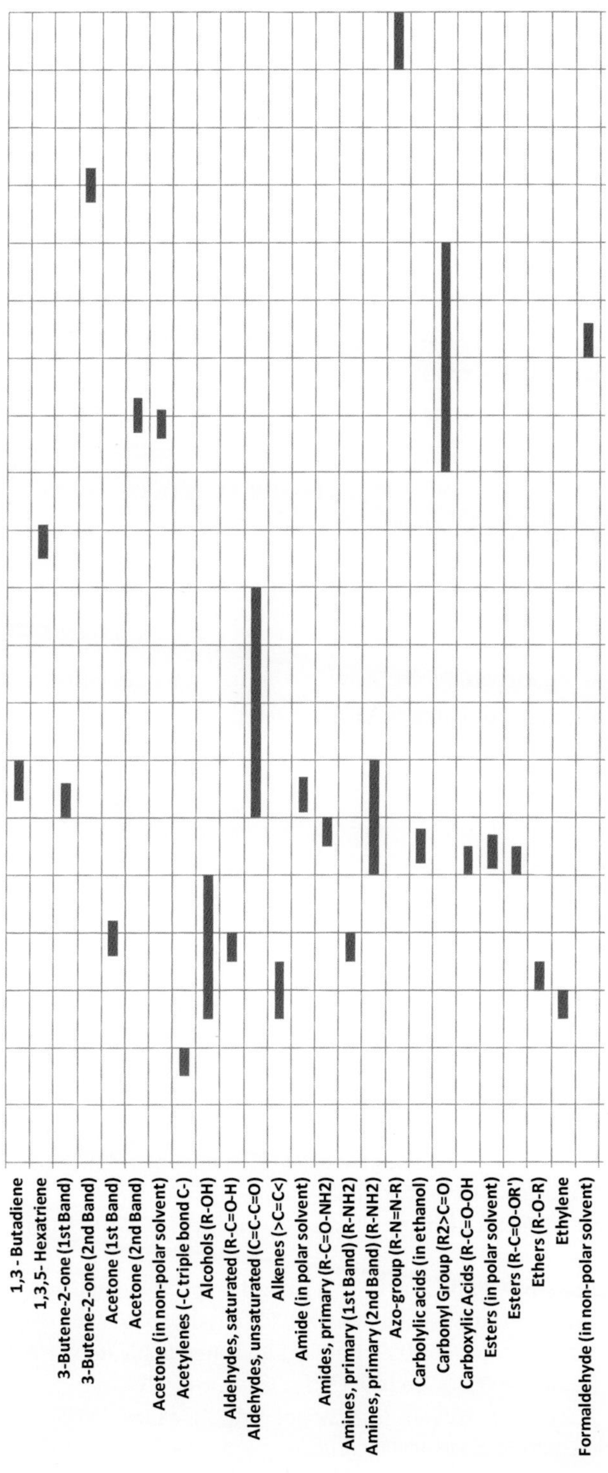

Fig. 7.9 Ultraviolet chromophore band positions in nanometers.

the comparative (synthetic or processed) proteins are not the same as the reference protein, and will therefore not be assumed to exhibit similar biological activity. Note the proportion of each secondary structure present is equal to the arithmetic sum of the fractional CD spectrum of each conformational reference protein spectrum.

7.7 Interpretation and Band Shifts in Ultraviolet Spectroscopy

Ultraviolet spectra follow different rules in terms of conjugation, intensity, and absorptivity, as compared to molecular spectroscopy, such as infrared and Raman. Interpreting ultraviolet spectra requires some knowledge of these rules for various applications. The basic rules are described in Table 7.8. as a summary of the band positions and intensities found for various compounds using UV measurements. Note that the extinction coefficient or absorptivity of various chemical moieties is given as Equation 7.7.

$$\varepsilon = \frac{A}{cl} \tag{7.7}$$

Where ε is the molar absorptivity (referred to as molar extinction coefficient by earlier physicists) in units of Liter $\cdot$ Mole^{-1} $\cdot$ cm^{-1}; c is the concentration of molecules in the spectrometer beam in units of Mole $\cdot$ Liter^{-1} (Note: this is a scaled volume fraction unit); and pathlength (l) is the thickness of the sample in units of cm of the measured sample at a specific concentration. The absorptivity for any specific molecule type is calculated by careful measurements of the absorbance of a compound, generally diluted in a suitable organic solvent, and by applying this equation. See the tables at the beginning of this chapter for more details on ultraviolet chromophore band positions.[24]

References

1. J.J. Workman and A.W. Springsteen (Eds.), *Applied Spectroscopy: A Compact Reference for Practitioners* (Academic Press, Boston, 1998).
2. American Conference of Governmental Industrial Hygienists (ACGIH) Ultraviolet Radiation Guidelines ©2010.
3. Guidelines for operators of Ultraviolet (UV) Tanning Lamps, in compliance with the provisions of Australia/New Zealand Standard AS/NZS 2635:2008 *Solaria for cosmetic purposes.* Tanning guidelines for operators Feb 2009; National Radiation Laboratory, 108 Victoria Street, Christchurch, New Zealand P O Box 25 099, Christchurch, New Zealand.
4. K.M. Wiggins, M. Hartung, O. Althoff, C. Wastian, and S.B. Mitra, Curing performance of a new-generation light-emitting diode dental curing unit, *J Am Dent Assoc.* **135**(10), 1471–1479.

5. Guidance for Industry and FDA Staff: Dental Curing Lights — Premarket Notification [510(k)], Document Issued on: March 27, 2006. U.S. Department of Health and Human Services Food and Drug Administration, Center for Devices and Radiological Health.
6. ASTM Document STP500, Ultraviolet Spectroscopy, ASTM International, West Conshohocken, PA, www.astm.org.
7. ASTM E2193-08, Standard Test Method for Ultraviolet Transmittance of Monoethylene Glycol (Ultraviolet Spectrophotometric Method) ASTM International, West Conshohocken, PA, 2008, www.astm.org.
8. MNL17-2ND, Ultraviolet/Visible Spectroscopy ASTM International, West Conshohocken, PA, 2014, www.astm.org.
9. S.K. Bramble, Fluorescence spectroscopy as an aid to imaging latent fingermarks in the ultraviolet, *Journal of Forensic Sciences* **41**(6) (1996), 1038–1041.
10. T.J. Siek, Identification of Drugs and Other Toxic Compounds from Their Ultraviolet Spectra. Part I: Ultraviolet Absorption Properties of Sixteen Structural Groups, *Journal of Forensic Sciences*, **19**(2) (1974), 193–214.
11. T.J. Siek and R.J. Osiewicz, Identification of drugs and other toxic compounds from their ultraviolet spectra. Part II: Ultraviolet absorption properties of thirteen structural groups, *Journal of Forensic Sciences* **20**(1) (1975), 18–37.
12. T.J. Siek, R.J. Osiewicz, and R.J. Bath, Identification of drugs and other toxic compounds from their ultraviolet spectra. Part III: Ultraviolet absorption properties of 22 structural groups, *Journal of Forensic Sciences* **21**(3) (1976), 525–551.
13. ASTM D2008-12, Standard Test Method for Ultraviolet Absorbance and Absorptivity of Petroleum Products, ASTM International, West Conshohocken, PA, 2012, www.astm.org.
14. ASTM D2269-10, Standard Test Method for Evaluation of White Mineral Oils by Ultraviolet Absorption ASTM International, West Conshohocken, PA, 2010, www.astm.org.
15. T. Bastow, W.H. Durnie, A. Jefferson, and J. Pang, Ultraviolet spectroscopy for the analysis of oil-in-water effluent using isopropanol as co-solvent, *Applied Spectroscopy* **51**(3) (1997), 318–322.
16. A.D. Skaja and S.G. Croll, Quantitative ultraviolet spectroscopy in weathering of a model polyester–urethane coating, *Polymer Degradation and Stability* **79**(1) (2003), 123–131.
17. K. Rosenheck and P. Doty, The far ultraviolet absorption spectra of polypeptide and protein solutions and their dependence on conformation, *Proceedings of the National Academy of Sciences of the United States of America* **47**(11) (1961), 1775.
18. D.B. Wetlaufer, Ultraviolet spectra of proteins and amino acids, *Adv. Protein Chem.* **17** (1962), 303–390.
19. J.T. Yang, C.-S.C. Wu, and H.M. Martinez, Calculation of protein conformation from circular dichroism, *Methods in Enzymology* **130** (1986), 208–269.
20. Protein Structures: Protein Data Bank image: http://www.rcsb.org/pdb/home/home.do
21. Circular dichroism using ultraviolet Theory: http://www.chem.uci.edu/~dmitryf/manuals/Fundamentals/CD%20spectroscopy.pdf
22. CD Method Described: http://www.ncbi.nlm.nih.gov/pmc/articles/PMC2728378/
23. Application of CD: http://www.ap-lab.com/circular_dichroism.htm
24. J. Workman, The Handbook of Organic Compounds, *Three-Volume Set: NIR, IR, Raman, and UV-Vis Spectra Featuring Polymers and Surfactants*, 1st edn (Elsevier-Academic Press, 2000).

Chapter 8

Ultraviolet Reference Material Spectra and Tables

Contents

8.1 Introduction to Standard Reference Materials (SRMs) for Photometric Accuracy Measurements

The National Institute for Standards and Technology (NIST) previously known as the National Bureau of Standards (NBS) provides materials designed as reference sources to verify the transmittance and absorbance scale accuracy of spectrophotometers. The materials are termed SRMs or Standard Reference Materials.[1]

8.1.1 *The Meaning of Photometric Accuracy*

Definition: Photometric accuracy is a formal term used to describe how well a measuring device (spectrophotometer or illuminance meter) is able to determine the total energy flux transmitted through (or reflected from) a Standard Reference Material (SRM). As stated in volume 14.01 of the American Society for Testing and Materials (ASTM) International,[2–3] the measurement of the photometric accuracy for an instrument is obtained as follows (Sections 20.1 and 20.3):

"Select the appropriate Standard Reference Material and obtain ten successive readings of the apparent absorbance or transmittance at the specified wavelength. Average the ten readings. The photometric accuracy is the difference between the true absorbance or transmittance value and the average observed value."

It is further stated that the following apply to the reporting of photometric accuracy.

"Report the photometric accuracy in the following order: reference material, wavelength, true absorbance or transmittance, observed absorbance or transmittance plus or minus the standard deviation."

8.1.2 *Example of Reporting Photometric Accuracy*

As an example the stated specifications for photometric accuracy for a test instrument might be written as:

Photometric Accuracy

±0.002 Au (for range of 0.0 to 0.5 Absorbance units)

±0.004 Au (for range of 0.5 to 1.0 Absorbance units)

±0.3%T

SRM: Measured with NIST 930D filters

Here is an evaluation of these specifications in greater detail. A photometric accuracy of "0.002 Au at 0 to 0.5 Au" indicates that after the instrument has been set to zero, a standard sample which transmits between 30.1%T (i.e., 0.5 Au) and 100%T (i.e., 0.0 Au) measures within ±0.002 Au as compared to the specified value for that SMR, Standard Reference Material at any wavelength position. Tables 8.1 and 8.2 are used to convert from transmittance or reflectance scale to absorbance scale. This issue of how to explain a specification such as ±0.004 at 0.5 to 1.0 Au, when using the NIST 930D filter set has been brought forward in competitive sales situations between instrument manufacturers. The tolerance for each of the 930D filters is stated by

Table 8.1 Relationship of %T and absorbance unit (Au) values.

T (as ppm)	%T	T	1/T	Au
10000000	1000	10	$0.1 = 10^{-1}$	−1.0
1000000	100	1	$1.0 = 10^{0}$	0.0
316230	31.6	0.316	3.16	0.5
100000	10.0	0.1	$10 = 10^{1}$	1.0
10000	1	0.01	$100 = 10^{2}$	2.0
1000	0.1	0.001	$1000 = 10^{3}$	3.0
100	0.01	0.0001	$10000 = 10^{4}$	4.0
10	0.001	0.00001	$100000 = 10^{5}$	5.0
1	0.0001	0.000001	$1000000 = 10^{6}$	6.0

Table 8.2 Expanded table showing the relationship of %T and %R to absorbance unit (Au) values.

Transmittance or Reflectance	Percent T or R	Absorbance (Au)
1.000	100.0	0.000000
0.995	99.5	0.002177
0.990	99.0	0.004365
0.985	98.5	0.006564
0.980	98.0	0.008774
0.975	97.5	0.010995
0.970	97.0	0.013228
0.965	96.5	0.015473
0.960	96.0	0.017729
0.955	95.5	0.019997
0.950	95.0	0.022276
0.945	94.5	0.024568
0.940	94.0	0.026872
0.935	93.5	0.029188
0.930	93.0	0.031517
0.925	92.5	0.033858
0.920	92.0	0.036212
0.915	91.5	0.038579
0.910	91.0	0.040959
0.905	90.5	0.043351
0.900	90.0	0.045757
0.895	89.5	0.048177
0.890	89.0	0.050610
0.885	88.5	0.053057
0.880	88.0	0.055517
0.875	87.5	0.057992
0.870	87.0	0.060481
0.865	86.5	0.062984

(*Continued*)

Table 8.2 (*Continued*).

Transmittance or Reflectance	Percent T or R	Absorbance (Au)
0.860	86.0	0.065502
0.855	85.5	0.068034
0.850	85.0	0.070581
0.845	84.5	0.073143
0.840	84.0	0.075721
0.835	83.5	0.078314
0.830	83.0	0.080922
0.825	82.5	0.083546
0.820	82.0	0.086186
0.815	81.5	0.088842
0.810	81.0	0.091515
0.805	80.5	0.094204
0.800	80.0	0.096910
0.795	79.5	0.099633
0.790	79.0	0.102373
0.785	78.5	0.105130
0.780	78.0	0.107905
0.775	77.5	0.110698
0.770	77.0	0.113509
0.765	76.5	0.116339
0.760	76.0	0.119186
0.755	75.5	0.122053
0.750	75.0	0.124939
0.745	74.5	0.127844
0.740	74.0	0.130768
0.735	73.5	0.133713
0.730	73.0	0.136677
0.725	72.5	0.139662
0.720	72.0	0.142668
0.715	71.5	0.145694
0.710	71.0	0.148742
0.705	70.5	0.151811
0.700	70.0	0.154902
0.695	69.5	0.158015
0.690	69.0	0.161151
0.685	68.5	0.164309
0.680	68.0	0.167491
0.675	67.5	0.170696
0.670	67.0	0.173925
0.665	66.5	0.177178
0.660	66.0	0.180456
0.655	65.5	0.183759
0.650	65.0	0.187087
0.645	64.5	0.190440
0.640	64.0	0.193820
0.635	63.5	0.197226

(*Continued*)

Table 8.2 (*Continued*).

Transmittance or Reflectance	Percent T or R	Absorbance (Au)
0.630	63.0	0.200659
0.625	62.5	0.204120
0.620	62.0	0.207608
0.615	61.5	0.211125
0.610	61.0	0.214670
0.605	60.5	0.218245
0.600	60.0	0.221849
0.595	59.5	0.225483
0.590	59.0	0.229148
0.585	58.5	0.232844
0.580	58.0	0.236572
0.575	57.5	0.240332
0.570	57.0	0.244125
0.565	56.5	0.247952
0.560	56.0	0.251812
0.555	55.5	0.255707
0.550	55.0	0.259637
0.545	54.5	0.263603
0.540	54.0	0.267606
0.535	53.5	0.271646
0.530	53.0	0.275724
0.525	52.5	0.279841
0.520	52.0	0.283997
0.515	51.5	0.288193
0.510	51.0	0.292430
0.505	50.5	0.296709
0.500	50.0	0.301030
0.495	49.5	0.305395
0.490	49.0	0.309804
0.485	48.5	0.314258
0.480	48.0	0.318759
0.475	47.5	0.323306
0.470	47.0	0.327902
0.465	46.5	0.332547
0.460	46.0	0.337242
0.455	45.5	0.341989
0.450	45.0	0.346787
0.445	44.5	0.351640
0.440	44.0	0.356547
0.435	43.5	0.361511
0.430	43.0	0.366532
0.425	42.5	0.371611
0.420	42.0	0.376751
0.415	41.5	0.381952
0.410	41.0	0.387216
0.405	40.5	0.392545

(*Continued*)

Table 8.2 (*Continued*).

Transmittance or Reflectance	Percent T or R	Absorbance (Au)
0.400	40.0	0.397940
0.395	39.5	0.403403
0.390	39.0	0.408935
0.385	38.5	0.414539
0.380	38.0	0.420216
0.375	37.5	0.425969
0.370	37.0	0.431798
0.365	36.5	0.437707
0.360	36.0	0.443697
0.355	35.5	0.449772
0.350	35.0	0.455932
0.345	34.5	0.462181
0.340	34.0	0.468521
0.335	33.5	0.474955
0.330	33.0	0.481486
0.325	32.5	0.488117
0.320	32.0	0.494850
0.315	31.5	0.501689
0.310	31.0	0.508638
0.305	30.5	0.515700
0.300	30.0	0.522879
0.295	29.5	0.530178
0.290	29.0	0.537602
0.285	28.5	0.545155
0.280	28.0	0.552842
0.275	27.5	0.560667
0.270	27.0	0.568636
0.265	26.5	0.576754
0.260	26.0	0.585027
0.255	25.5	0.593460
0.250	25.0	0.602060
0.245	24.5	0.610834
0.240	24.0	0.619789
0.235	23.5	0.628932
0.230	23.0	0.638272
0.225	22.5	0.647817
0.220	22.0	0.657577
0.215	21.5	0.667562
0.210	21.0	0.677781
0.205	20.5	0.688246
0.200	20.0	0.698970
0.195	19.5	0.709965
0.190	19.0	0.721246
0.185	18.5	0.732828
0.180	18.0	0.744727
0.175	17.5	0.756962

(*Continued*)

Table 8.2 (*Continued*).

Transmittance or Reflectance	Percent T or R	Absorbance (Au)
0.170	17.0	0.769551
0.165	16.5	0.782516
0.160	16.0	0.795880
0.155	15.5	0.809668
0.150	15.0	0.823909
0.145	14.5	0.838632
0.140	14.0	0.853872
0.135	13.5	0.869666
0.130	13.0	0.886057
0.125	12.5	0.903090
0.120	12.0	0.920819
0.115	11.5	0.939302
0.110	11.0	0.958607
0.105	10.5	0.978811
0.100	10.0	1.000000
0.095	9.50	1.022276
0.090	9.00	1.045757
0.085	8.50	1.070581
0.080	8.00	1.096910
0.075	7.50	1.124939
0.070	7.00	1.154902
0.065	6.50	1.187087
0.060	6.00	1.221849
0.055	5.50	1.259637
0.050	5.00	1.301030
0.045	4.50	1.346787
0.040	4.00	1.397940
0.035	3.50	1.455932
0.030	3.00	1.522879
0.025	2.50	1.602060
0.020	2.00	1.698970
0.015	1.50	1.823909
0.010	1.00	2.000000
0.009	0.90	2.045757
0.008	0.80	2.096910
0.007	0.70	2.154902
0.006	0.60	2.221849
0.005	0.50	2.301030
0.004	0.40	2.397940
0.003	0.30	2.522879
0.002000	0.2000	2.698970
0.001500	0.1500	2.823909
0.001000	0.1000	3.000000
0.000100	0.0100	4.000000
0.000010	0.0010	5.000000
0.000001	0.0001	6.000000

NIST as ±0.5 percent relative. Thus, a 10%T filter (which corresponds to 0.1 T or 1.0 Au) neutral density 930D-type reference standard would be expected to be within the range of 9.95 to 10.05%T. These values are equivalent to 1.0022 Au to 0.9978 Au. The difference in Au is 1.0022 Au minus 0.9978 Au equals 0.0044 Au. The standard tolerance in Au is 1.0000 ± 0.0022 Au, rounded to 1.000 ± 0.002 Au. Conservatively for transmittance we state 10.00 ± 0.03%T; and related to 0.0 Au or 100%T, we specify 100.0 ±0.3%T.

If the stated tolerance for the NIST 930D filter is ±0.002 Au at 1.0 Au, we are stating that our maximum deviation in photometric accuracy for our instrument at 1.0 Au is also ± 0.002 Au. For the worst case measurement for the NIST filter, we would have the filter error of ±0.002 added to the instrument error of ±0.002, to result in a total maximum variation ±0.004 Au error (this is our stated specification for the example given).

And one can readily calculate the reflectance or transmittance of any measurement given the absorbance of that measurement as follows. For conversion of Absorbance to Transmittance we use:

$$A = -\log_{10}\left(\frac{I}{I_0}\right) = \log_{10}\left(\frac{I_0}{I}\right) \tag{8.1}$$

$$\frac{I}{I_0} = 10^{-A} \Rightarrow \frac{I_0}{I} = \frac{1}{10^{-A}} \tag{8.2}$$

$$\frac{I}{I_0} = 10^{-A} = \mathrm{T} \tag{8.3}$$

$$\%\mathrm{T} = \mathrm{T} \times 100 \tag{8.4}$$

To convert Absorbance to Reflectance we use:

$$A = -\log_{10}\left(\frac{I}{I_0}\right) = \log_{10}\left(\frac{I_0}{I}\right) \tag{8.5}$$

$$\frac{I}{I_0} = 10^{-A} \Rightarrow \frac{I_0}{I} = \frac{1}{10^{-A}} \tag{8.6}$$

$$\frac{I}{I_0} = 10^{-A} = \mathrm{R} \tag{8.7}$$

$$\%\mathrm{R} = \mathrm{R} \times 100 \tag{8.8}$$

8.2 Ultraviolet (UV) Wavelength Measurement Standards

Reference spectra and data provided with permission by Starna Ltd., Hainault, IG6 3UT, UK. www.starna.com.[4] Note these standards are also useful and traceable for Visible (Vis) spectroscopic measurements. The certified peak positions are labeled within the figures and the data are given in accompanying tables.

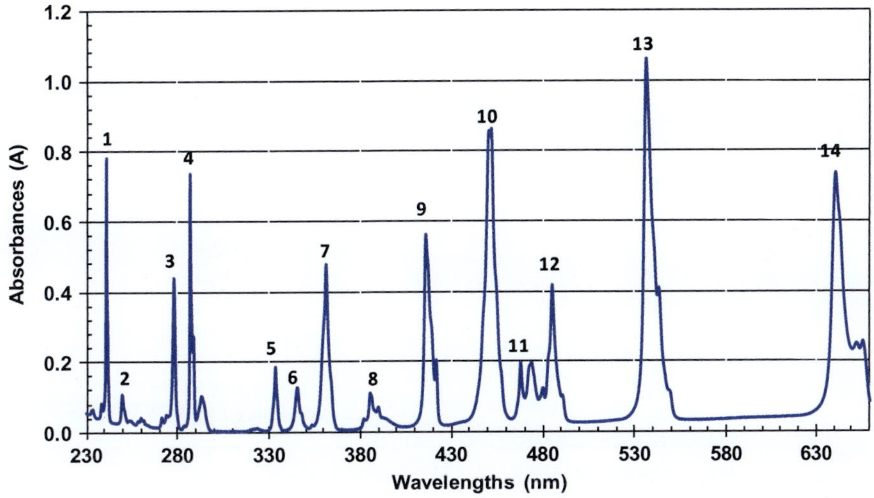

Fig. 8.1a High resolution absorbance spectrum of holmium oxide liquid wavelength standard traceable to SRM 2034 (Data and spectrum provided with permission by Starna Ltd., Hainault, IG6 3UT, UK. www.starna.com).

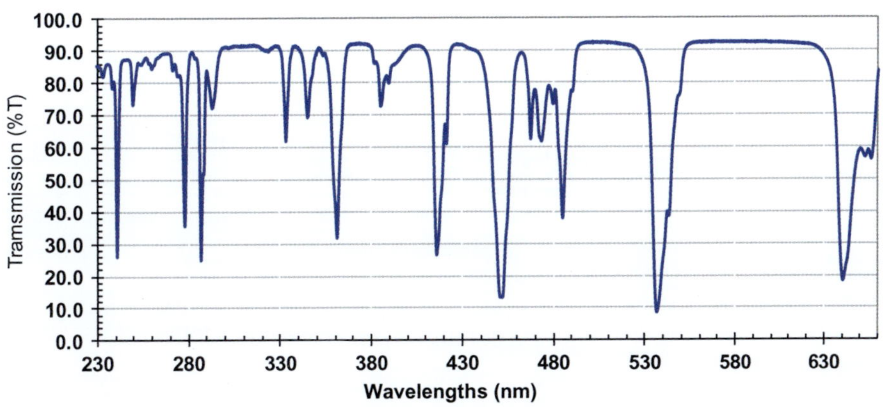

Fig. 8.1b High resolution transmittance spectrum of holmium oxide liquid wavelength standard traceable to SRM 2034 (Data and spectrum provided with permission by Starna Ltd., Hainault, IG6 3UT, UK. www.starna.com).

Table 8.3 Data table of certified wavelengths for absorbance spectrum of holmium oxide liquid wavelength standard traceable to SRM 2034. Consists of holmium Oxide (4%) in perchloric acid (10%) sealed in a quartz cell. (Data and spectrum provided with permission by Starna Ltd., Hainault, IG6 3UT, UK. www.starna.com).

Designation number	Certified wavelengths
1	241.0
2	249.9
3	278.2
4	287.1
5	333.5
6	345.6
7	361.4
8	385.4
9	416.1
10	451.1
11	467.8
12	485.2
13	536.5
14	640.6

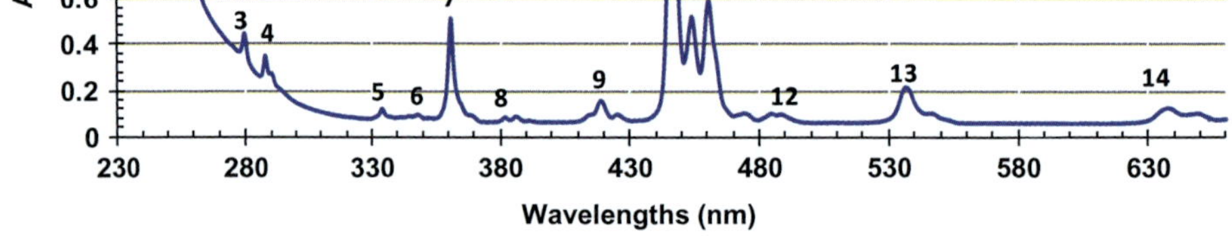

Fig. 8.2a Absorbance spectrum of holmium oxide glass wavelength standard traceable to SRM 2034 (Data and spectrum provided with permission by Starna Ltd., Hainault, IG6 3UT, UK. www.starna.com).

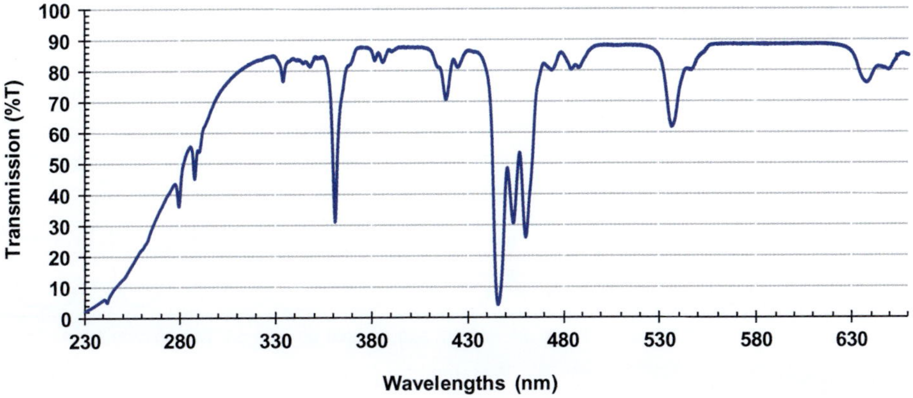

Fig. 8.2b Transmittance spectrum of holmium oxide glass wavelength standard traceable to SRM 2034 (Data and spectrum provided with permission by Starna Ltd., Hainault, IG6 3UT, UK. www.starna.com).

Table 8.4 Data table of certified wavelengths for absorbance spectrum of holmium oxide glass filter wavelength standard traceable to SRM 2034. Consists of holmium oxide in glass. (Data and spectrum provided with permission by Starna Ltd., Hainault, IG6 3UT, UK. www.starna.com).

Designation number	Certified wavelengths (nm)
1	241.0
None	249.9
3	278.2
4	287.1
5	333.5
6	345.6
7	361.4
8	385.4
9	416.1
10	451.1
None	467.8
Questionable	485.2
13	536.5
14	640.6

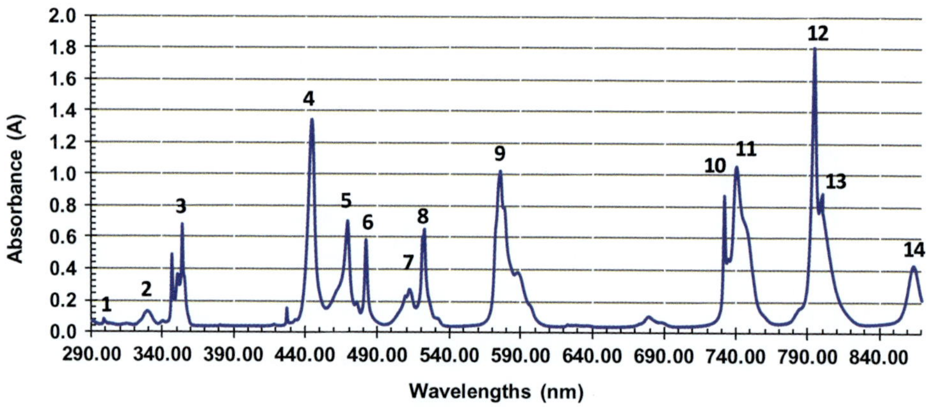

Fig. 8.3a Absorbance spectrum of didymium oxide liquid wavelength standard traceable to SRM 2034 (Data and spectrum provided with permission by Starna Ltd., Hainault, IG6 3UT, UK. www.starna.com).

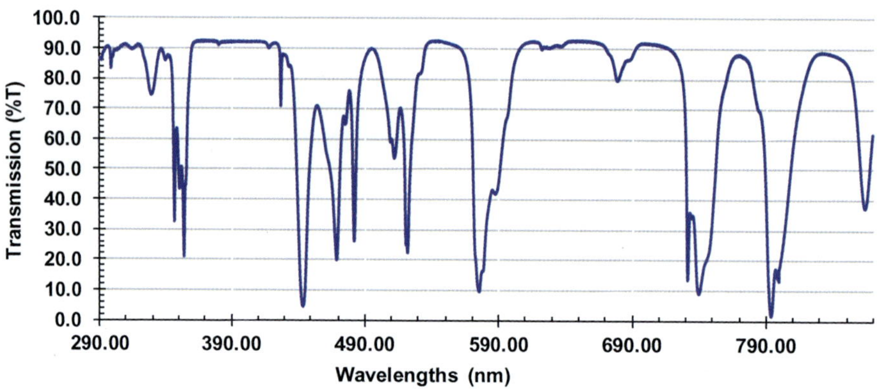

Fig. 8.3b Transmittance spectrum of didymium oxide liquid wavelength standard traceable to SRM 2034 (Data and spectrum provided with permission by Starna Ltd., Hainault, IG6 3UT, UK. www.starna.com).

Table 8.5 Data table absorbance spectrum of didymium oxide liquid wavelength standard traceable to SRM 2034. Didymium oxide in perchloric acid, sealed in a quartz cell. (Data and spectrum provided with permission by Starna Ltd., Hainault, IG6 3UT, UK. www.starna.com).

Designation number	Certified Wavelength (nm)
1	298.0
2	329.0
3	354.0
4	444.1
5	468.7
6	481.8
7	512.0
8	521.3
9	574.9
10	731.8
11	740.1
12	794.1
13	801.2
14	865.0

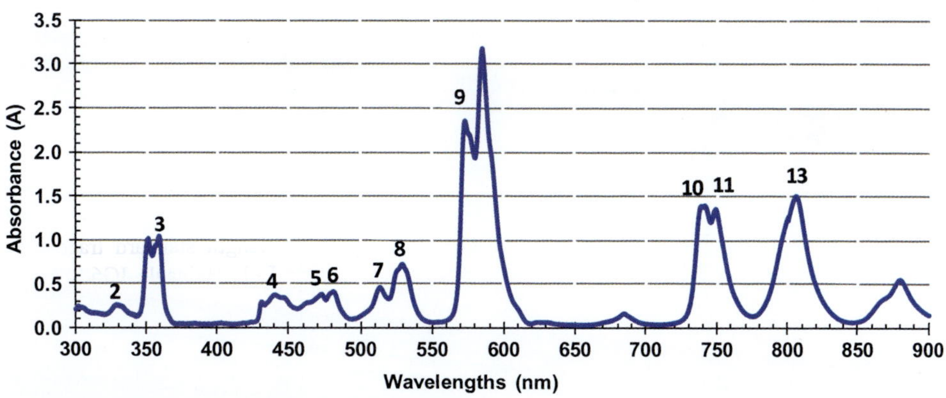

Fig. 8.4a Absorbance spectrum of didymium oxide glass wavelength standard traceable to SRM 2034 (Data and spectrum provided with permission by Starna Ltd., Hainault, IG6 3UT, UK. www.starna.com).

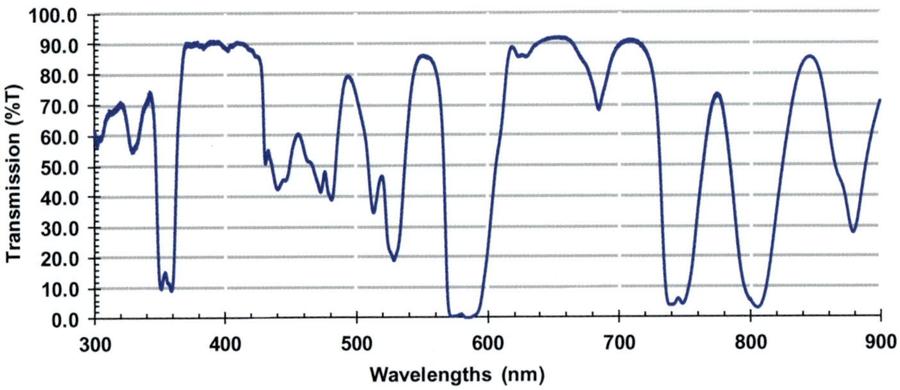

Fig. 8.4b Transmittance spectrum of didymium oxide glass wavelength standard traceable to SRM 2034 (Data and spectrum provided with permission by Starna Ltd., Hainault, IG6 3UT, UK. www.starna.com).

Table 8.6 Data table for absorbance spectrum of didymium oxide glass wavelength standard traceable to SRM 2034. Didymium oxide in glass. (Data and spectrum provided with permission by Starna Ltd., Hainault, IG6 3UT, UK. www.starna.com).

Designation number	Certified wavelength (nm)
None	298.0
2	329.0
3	354.0
4	444.1
5	468.7
6	481.8
7	512.0
8	521.3
9	574.9
10	731.8
11	740.1
None	794.1
13	801.2
None	865.0

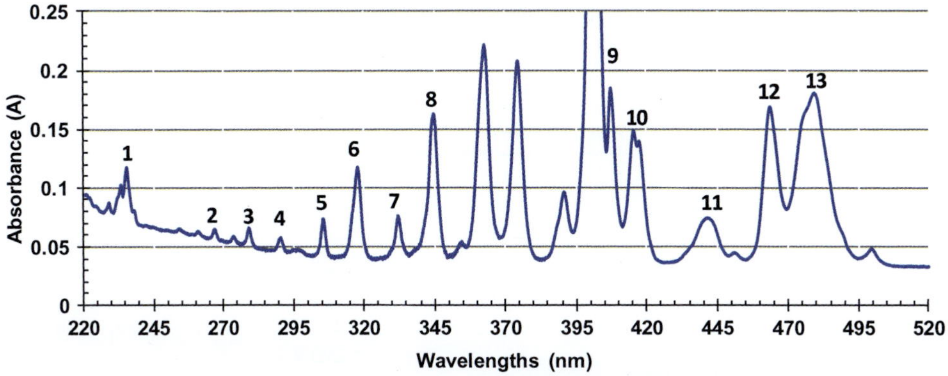

Fig. 8.5a Absorbance spectrum of samarium oxide wavelength standard traceable to SRM 2034 (Data and spectrum provided with permission by Starna Ltd., Hainault, IG6 3UT, UK. www.starna.com).

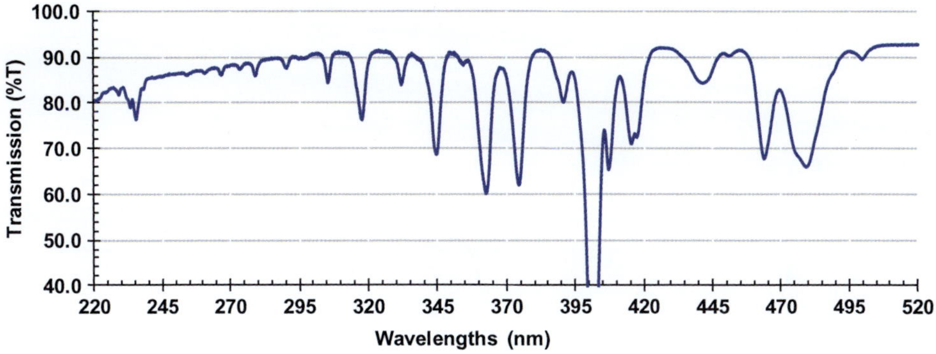

Fig. 8.5b Transmittance spectrum of samarium oxide wavelength standard traceable to SRM 2034 (Data and spectrum provided with permission by Starna Ltd., Hainault, IG6 3UT, UK. www.starna.com).

Table 8.7 Data table for absorbance spectrum of samarium oxide wavelength standard traceable to SRM 2034 (Data and spectrum provided with permission by Starna Ltd., Hainault, IG6 3UT, UK. www.starna.com).

Designation number	Certified wavelength (nm)
1	235.2
2	266.7
3	279.1
4	290.2
5	305.3
6	317.6
7	331.7
8	344.6
9	401.5
10	415.2
11	441.2
12	463.5
13	478.5

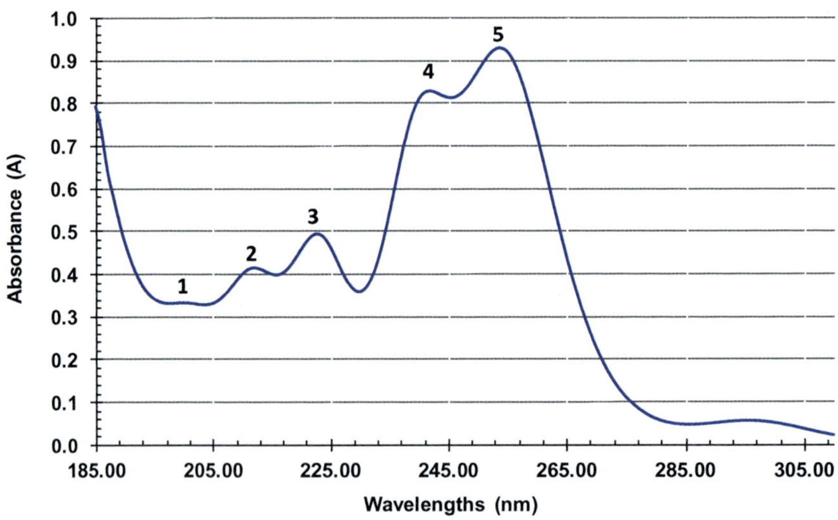

Fig. 8.6a Absorbance spectrum of rare earth oxide wavelength standard traceable to SRM 2034 (Data and spectrum provided with permission by Starna Ltd., Hainault, IG6 3UT, UK. www.starna.com).

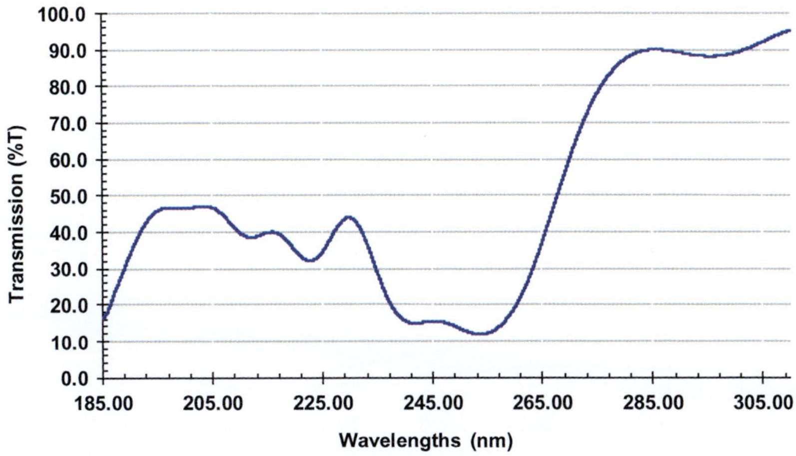

Fig. 8.6b Transmittance spectrum of rare earth oxide wavelength standard traceable to SRM 2034 (Data and spectrum provided with permission by Starna Ltd., Hainault, IG6 3UT, UK. www.starna.com).

Table 8.8 Data table for absorbance spectrum of rare earth oxide wavelength standard traceable to SRM 2034 (Data and spectrum provided with permission by Starna Ltd., Hainault, IG6 3UT, UK. www.starna.com).

Designation number	Certified wavelength (nm)
1	201.0
2	212.0
3	223.0
4	240.0
5	253.0

8.3 Ultraviolet Photometric (Absorbance) Measurement Standards

Reference spectra and data provided with permission by Starna Ltd., Hainault, IG6 3UT, UK. www.starna.com. Note these standards are also useful and traceable for Visible (Vis) spectroscopic measurements.

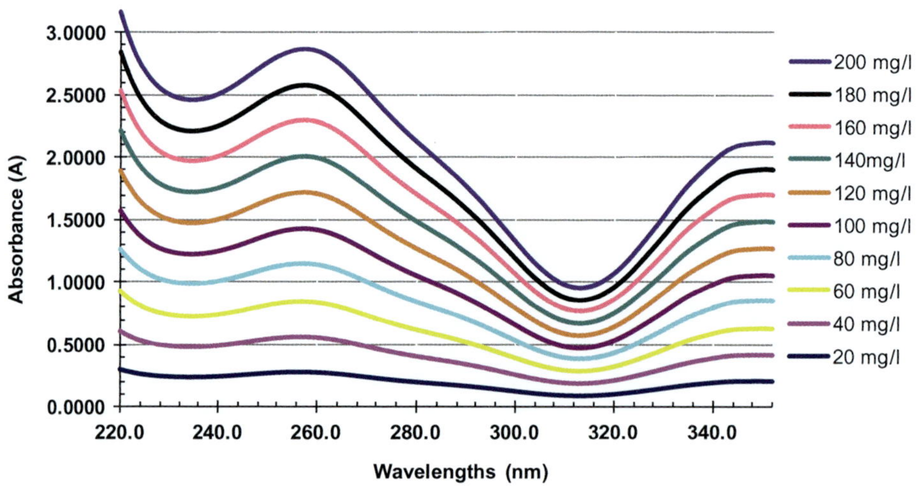

Fig. 8.7a Potassium dichromate absorbance and linearity standard traceable to SRM 935a. (Data and spectrum provided with permission by Starna Ltd., Hainault, IG6 3UT, UK. www.starna.com).

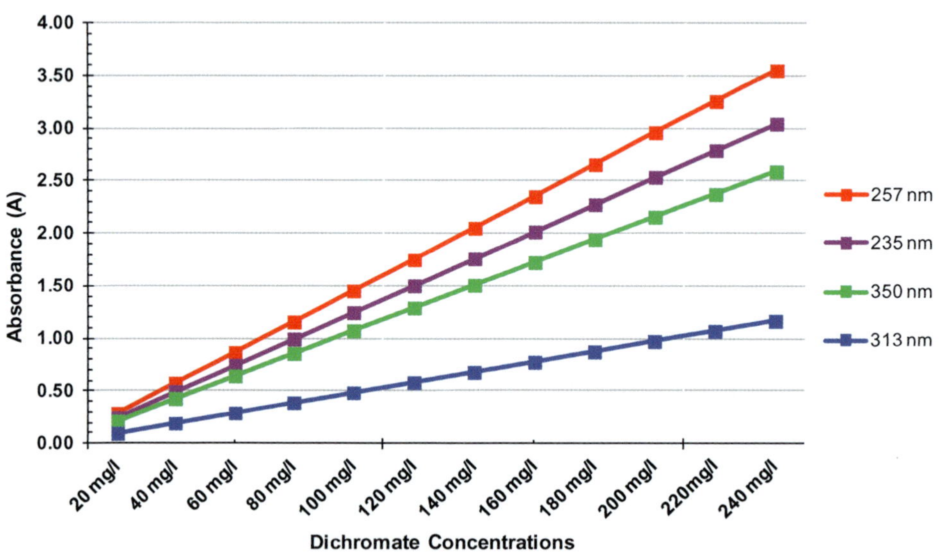

Fig. 8.7b Potassium chromate absorbance linearity as a photometric standard using five concentrations of standard. This is a test for photometric linearity throughout the ultraviolet region. (Data and spectrum provided with permission by Starna Ltd., Hainault, IG6 3UT, UK. www.starna.com).

Table 8.9 Data table for potassium dichromate absorbance and linearity standard traceable to SRM 935a. Values shown are the absorbance values for each measured wavelength at the designated concentration of potassium dichromate. (Data and spectrum provided with permission by Starna Ltd., Hainault, IG6 3UT, UK. www.starna.com).

Wavelength (nm)	20 mg/l Abs	40 mg/l Abs	60 mg/l Abs	80 mg/l Abs	100 mg/l Abs	120 mg/l Abs	140 mg/l Abs	160 mg/l Abs	180 mg/l Abs	200 mg/l Abs
350.0	0.2104	0.4137	0.6289	0.8497	1.0591	1.2713	1.4834	1.6956	1.9078	2.1199
313.0	0.0948	0.1858	0.2834	0.3827	0.4764	0.5718	0.6673	0.7627	0.8581	0.9535
257.0	0.2807	0.5550	0.8469	1.1468	1.4303	1.7177	2.0051	2.2925	2.5799	2.8673
235.0	0.2404	0.4766	0.7280	0.9860	1.2271	1.4738	1.7205	1.9672	2.2139	2.4605

Table 8.10 Data table for nicotinic acid absorbance standard traceable to SRM 935a. Values shown are the absorbance values for each measured wavelength at the designated concentration of nicotinic acid. (Data and spectrum provided with permission by Starna Ltd., Hainault, IG6 3UT, UK. www.starna.com).

Wavelength (nm)	6 mg/l Abs	12 mg/l Abs	18 mg/l Abs	24 mg/l Abs
261.0	0.281	0.572	0.862	1.159
213.0	0.243	0.492	0.741	0.996

Table 8.11 Data table for metal on quartz filters used for absorbance and linearity standard traceable to SRM 2031. Both absorbance and percent transmittance are given in the table. (Data and spectrum provided with permission by Starna Ltd., Hainault, IG6 3UT, UK. www.starna.com).

nm	Blank Abs.	1%T Abs.	10%T Abs.	30%T Abs.	50%T Abs.	90%T Abs.	nm	1%T %Trans.	10%T %Trans.	30%T %Trans.	50%T %Trans.	90%T %Trans.
200	0.0003	2.3447	1.0315	0.4456	0.3055	0.0580	200	0.4525	9.3058	35.8637	49.5130	87.5513
210	0.0002	2.3625	1.0384	0.4433	0.3012	0.0483	210	0.4342	9.1583	36.0456	50.0063	89.5160
220	0.0003	2.3610	1.0373	0.4342	0.2972	0.0461	220	0.4357	9.1835	36.8152	50.4692	89.9937
230	0.0001	2.3348	1.0268	0.4231	0.2926	0.0435	230	0.4628	9.4053	37.7579	50.9936	90.4988
240	0.0001	2.2888	1.0097	0.4178	0.2890	0.0414	240	0.5145	9.7830	38.2222	51.4171	90.9344
250	0.0002	2.2332	0.9892	0.4195	0.2866	0.0403	250	0.5848	10.2562	38.0812	51.7112	91.1789
260	0.0002	2.1767	0.9690	0.4263	0.2853	0.0396	260	0.6661	10.7433	37.4826	51.8711	91.3234
270	0.0002	2.1248	0.9514	0.4359	0.2844	0.0387	270	0.7504	11.1879	36.6644	51.9666	91.5042
280	0.0001	2.0807	0.9378	0.4470	0.2844	0.0379	280	0.8308	11.5445	35.7393	51.9735	91.6706
290	0.0002	2.0451	0.9284	0.4588	0.2849	0.0373	290	0.9017	11.7973	34.7892	51.9166	91.8260
300	0.0001	2.0183	0.9230	0.4709	0.2860	0.0368	300	0.9589	11.9413	33.8201	51.7681	91.9006
310	0.0002	1.9993	0.9209	0.4829	0.2875	0.0364	310	1.0021	12.0027	32.9131	51.6058	92.0110
320	0.0001	1.9868	0.9214	0.4946	0.2893	0.0360	320	1.0313	11.9883	32.0259	51.3794	92.0787
330	0.0002	1.9799	0.9241	0.5059	0.2914	0.0356	330	1.0478	11.9151	31.2091	51.1495	92.1706
340	0.0002	1.9773	0.9286	0.5166	0.2938	0.0354	340	1.0540	11.7927	30.4468	50.8569	92.1997
350	0.0003	1.9771	0.9338	0.5257	0.2958	0.0349	350	1.0548	11.6532	29.8265	50.6412	92.3420
360	0.0002	1.9791	0.9400	0.5344	0.2981	0.0344	360	1.0497	11.4855	29.2298	50.3593	92.4218
370	0.0001	1.9829	0.9469	0.5424	0.3004	0.0340	370	1.0403	11.3013	28.6884	50.0777	92.4928
380	0.0001	1.9880	0.9547	0.5505	0.3032	0.0340	380	1.0283	11.1024	28.1560	49.7587	92.4944
390	0.0001	1.9929	0.9623	0.5586	0.3056	0.0337	390	1.0166	10.9093	27.6368	49.4862	92.5371
400	0.0001	1.9980	0.9699	0.5670	0.3078	0.0336	400	1.0048	10.7203	27.1078	49.2333	92.5775
410	0.0001	2.0028	0.9774	0.5757	0.3099	0.0333	410	0.9938	10.5369	26.5693	49.0030	92.6303
420	0.0001	2.0073	0.9848	0.5848	0.3118	0.0332	420	0.9835	10.3579	26.0173	48.7760	92.6416
430	0.0000	2.0111	0.9918	0.5937	0.3135	0.0330	430	0.9748	10.1924	25.4861	48.5907	92.6895
440	0.0001	2.0146	0.9986	0.6025	0.3150	0.0329	440	0.9672	10.0354	24.9816	48.4339	92.7352
450	0.0001	2.0172	1.0050	0.6109	0.3162	0.0327	450	0.9614	9.8896	24.5066	48.2949	92.7684

(Continued)

Table 8.11 (*Continued*).

nm	Blank Abs.	1%T Abs.	10%T Abs.	30%T Abs.	50%T Abs.	90%T Abs.	nm	1%T %Trans.	10%T %Trans.	30%T %Trans.	50%T %Trans.	90%T %Trans.
460	0.0001	2.0197	1.0111	0.6185	0.3174	0.0327	460	0.9558	9.7493	24.0746	48.1536	92.7643
470	0.0001	2.0210	1.0168	0.6252	0.3183	0.0326	470	0.9529	9.6219	23.7090	48.0621	92.7945
480	0.0000	2.0221	1.0222	0.6309	0.3191	0.0325	480	0.9504	9.5030	23.3985	47.9695	92.7895
490	0.0001	2.0224	1.0272	0.6354	0.3197	0.0325	490	0.9500	9.3956	23.1546	47.9090	92.8119
500	0.0002	2.0219	1.0317	0.6387	0.3200	0.0323	500	0.9513	9.3013	22.9872	47.8882	92.8661
510	0.0001	2.0211	1.0359	0.6411	0.3203	0.0323	510	0.9528	9.2086	22.8562	47.8436	92.8485
520	0.0001	2.0198	1.0398	0.6424	0.3204	0.0323	520	0.9556	9.1257	22.7866	47.8279	92.8496
530	0.0001	2.0179	1.0433	0.6427	0.3203	0.0323	530	0.9599	9.0532	22.7696	47.8342	92.8606
540	0.0000	2.0157	1.0466	0.6423	0.3203	0.0323	540	0.9646	8.9840	22.7883	47.8357	92.8476
550	0.0001	2.0126	1.0493	0.6411	0.3200	0.0321	550	0.9716	8.9284	22.8539	47.8754	92.8949
560	0.0000	2.0095	1.0519	0.6394	0.3197	0.0320	560	0.9783	8.8744	22.9413	47.9047	92.9048
570	0.0001	2.0061	1.0541	0.6372	0.3192	0.0318	570	0.9863	8.8291	23.0607	47.9633	92.9438
580	0.0001	2.0025	1.0564	0.6348	0.3188	0.0318	580	0.9946	8.7852	23.1881	48.0086	92.9533
590	0.0001	1.9986	1.0582	0.6321	0.3182	0.0317	590	1.0035	8.7473	23.3346	48.0713	92.9875
600	0.0001	1.9944	1.0599	0.6292	0.3177	0.0316	600	1.0132	8.7129	23.4905	48.1336	93.0078
610	0.0001	1.9902	1.0616	0.6263	0.3172	0.0316	610	1.0230	8.6795	23.6495	48.1852	93.0046
620	0.0001	1.9860	1.0631	0.6232	0.3166	0.0315	620	1.0332	8.6509	23.8212	48.2538	93.0278
630	0.0002	1.9816	1.0645	0.6201	0.3161	0.0315	630	1.0437	8.6231	23.9913	48.3138	93.0349
640	0.0002	1.9770	1.0657	0.6169	0.3155	0.0315	640	1.0547	8.6004	24.1709	48.3846	93.0460
650	0.0002	1.9727	1.0669	0.6138	0.3149	0.0315	650	1.0653	8.5759	24.3426	48.4458	93.0399
660	0.0002	1.9682	1.0681	0.6107	0.3144	0.0315	660	1.0765	8.5541	24.5195	48.5091	93.0500
670	0.0003	1.9638	1.0691	0.6076	0.3139	0.0315	670	1.0875	8.5332	24.6953	48.5679	93.0552
680	0.0004	1.9596	1.0703	0.6046	0.3133	0.0315	680	1.0985	8.5130	24.8726	48.6464	93.0759
690	0.0004	1.9553	1.0714	0.6017	0.3129	0.0316	690	1.1093	8.4910	25.0398	48.6952	93.0683
700	0.0005	1.9512	1.0726	0.5990	0.3125	0.0318	700	1.1202	8.4705	25.2050	48.7504	93.0485

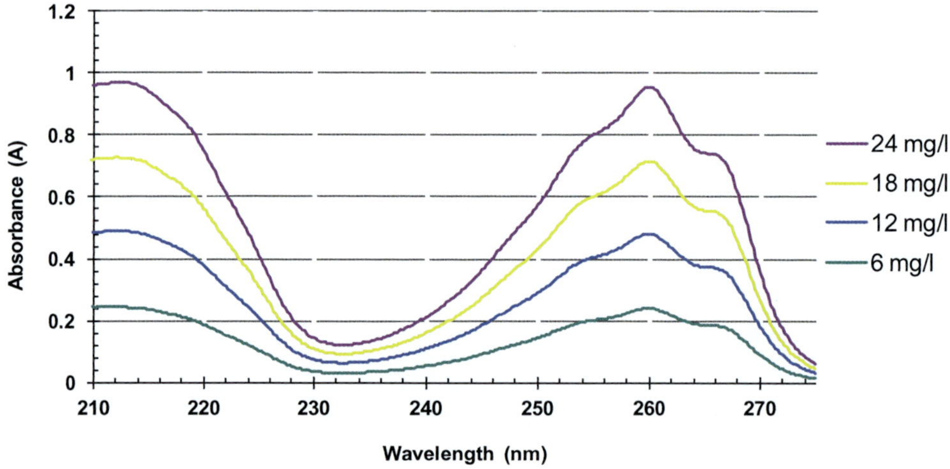

Fig. 8.8a Nicotinic acid absorbance standard traceable to SRM 935a (Data and spectrum provided with permission by Starna Ltd., Hainault, IG6 3UT, UK. www.starna.com).

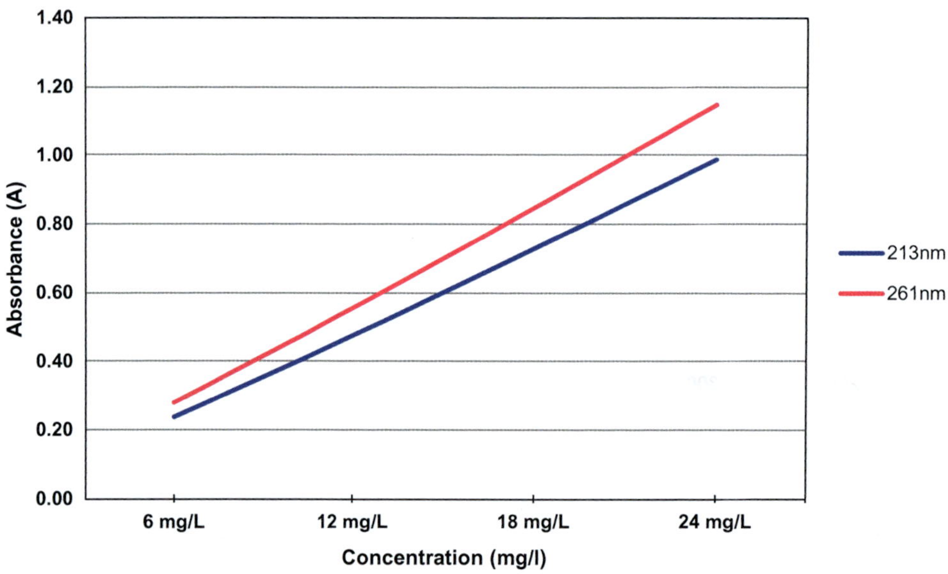

Fig. 8.8b Nicotinic acid absorbance linearity as a photometric standard using four concentrations of standard. This is a test for photometric linearity throughout the ultraviolet region. (Data and spectrum provided with permission by Starna Ltd., Hainault, IG6 3UT, UK. www.starna.com).

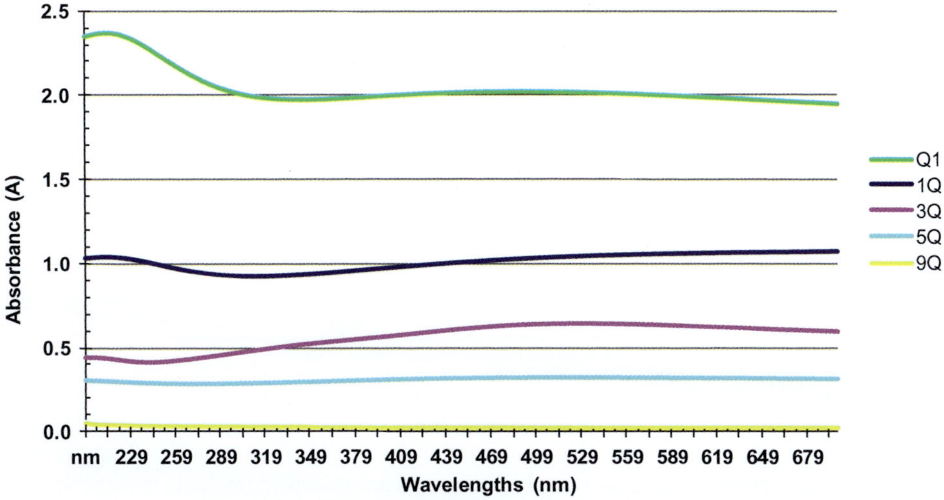

Fig. 8.9 Metal on quartz filters used for absorbance and linearity standard traceable to SRM 2031 (Data and spectrum provided with permission by Starna Ltd., Hainault, IG6 3UT, UK. www.starna.com).

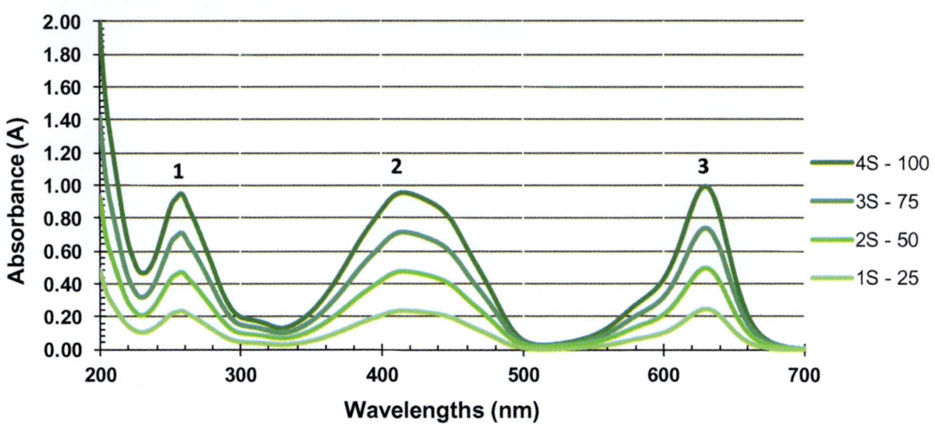

Fig. 8.10a Starna green absorbance, wavelength, and linearity standard traceable to SRM 935a and SRM 2034 (Data and spectrum provided with permission by Starna Ltd., Hainault, IG6 3UT, UK. www.starna.com). Certified wavelength peaks marked.

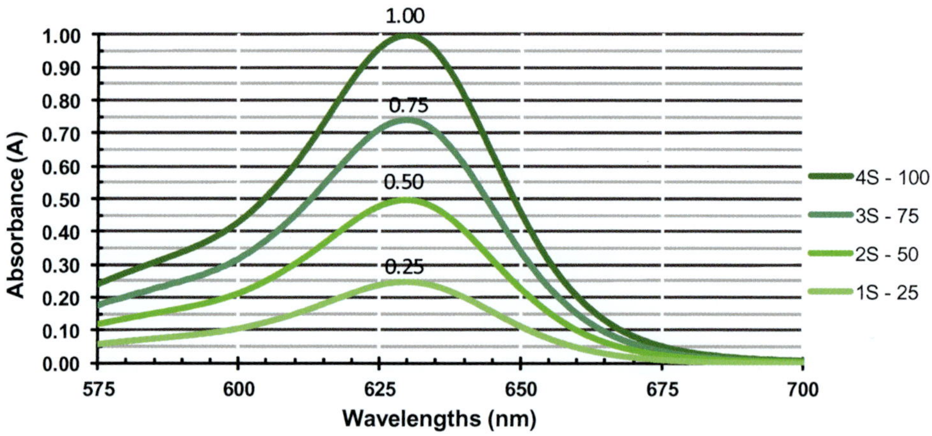

Fig. 8.10b Starna green absorbance, wavelength, and linearity standard traceable to SRM 935a and SRM 2034 (Data and spectrum provided with permission by Starna Ltd., Hainault, IG6 3UT, UK. www.starna.com). Certified absorbance bands marked at 630.0 nm.

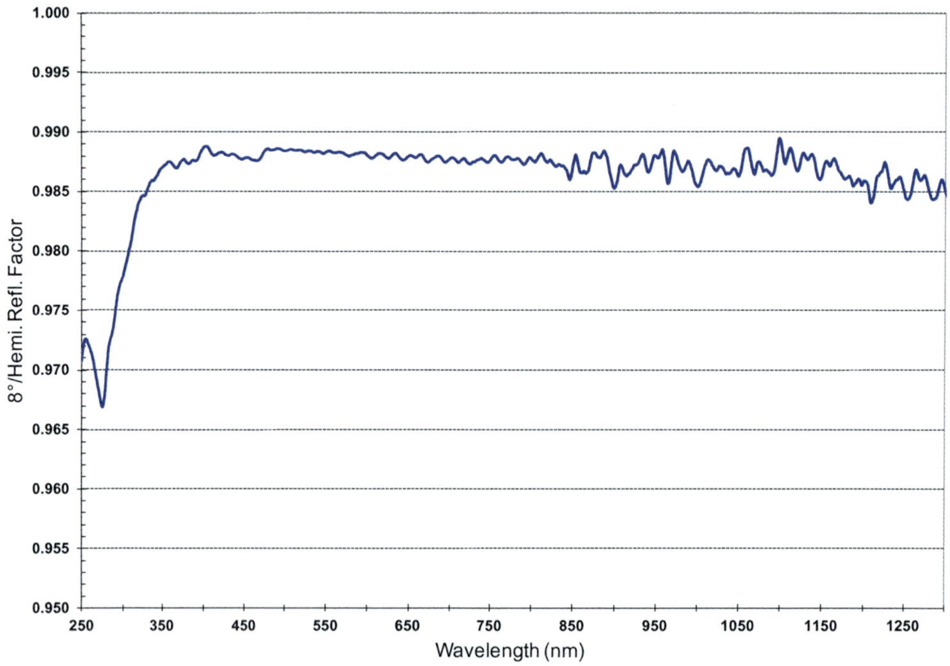

Fig. 8.11a Fluorilon FW99-02 reflectance wavelength standard measured using 8°/hemispherical spectral reflectance factor geometry and a calibrated double-monochromator instrument from 250 nm to 1300 nm (Data and spectrum used with permission from Avian Technologies, New London, NH).

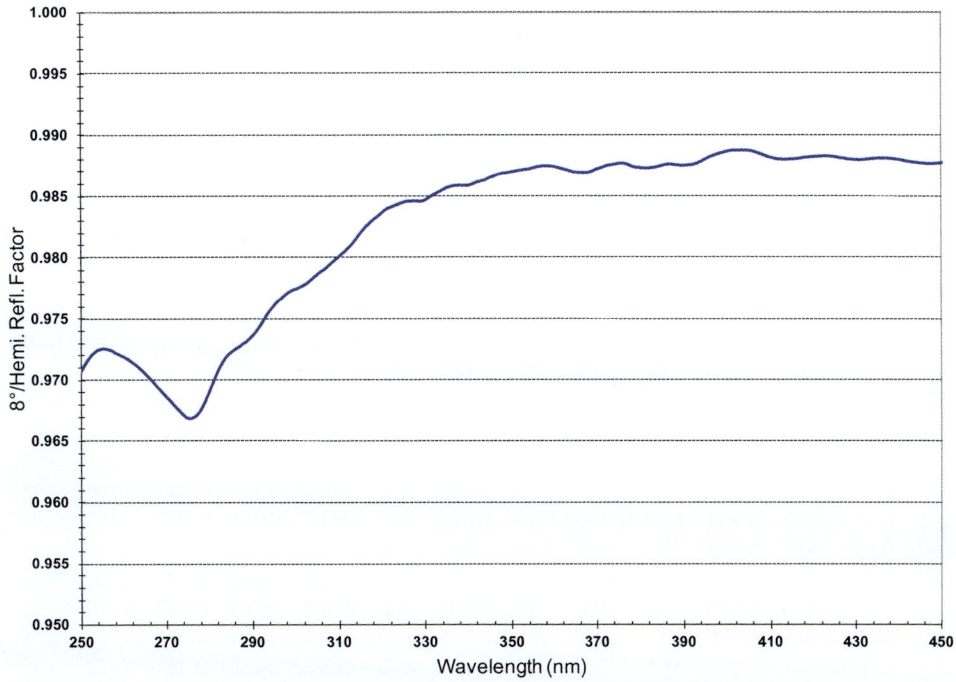

Fig. 8.11b Fluorilon FW99-02 reflectance wavelength standard measured using 8°/hemispherical spectral reflectance factor geometry and a calibrated double-monochromator instrument from 250 nm to 450 nm (Data and spectrum used with permission from Avian Technologies, New London, NH).

Table 8.12 Data table for starna green absorbance, wavelength, and linearity standard traceable to SRM 935a and SRM 2034 (Data and spectrum provided with permission by Starna Ltd., Hainault, IG6 3UT, UK. www.starna.com).

Designation number	Wavelengths certified (nm)	Absorbance certified (A) at 630 nm
1	257.0	0.25
2	416.0	0.50
3	630.0	0.75
		1.00

Table 8.13 Data table for fluorilon FW99-02 reflectance (R) wavelength (nm) standard measured using 8°/hemispherical spectral reflectance factor geometry and a calibrated double-monochromator instrument from 250 nm to 2500 nm (Data and spectrum used with permission from Avian Technologies, New London, NH).

nm	R	nm	R	nm	R	nm	R	nm	R	nm	R	nm	R
250	0.971	401	0.989	552	0.988	703	0.988	854	0.988	1005	0.986	1156	0.988
251	0.971	402	0.989	553	0.988	704	0.988	855	0.987	1006	0.987	1157	0.987
252	0.972	403	0.989	554	0.988	705	0.988	856	0.987	1007	0.987	1158	0.987
253	0.972	404	0.989	555	0.988	706	0.988	857	0.987	1008	0.987	1159	0.987
254	0.972	405	0.989	556	0.988	707	0.988	858	0.987	1009	0.987	1160	0.987
255	0.973	406	0.989	557	0.988	708	0.988	859	0.987	1010	0.988	1161	0.987
256	0.973	407	0.989	558	0.988	709	0.988	860	0.987	1011	0.988	1162	0.988
257	0.972	408	0.988	559	0.988	710	0.987	861	0.987	1012	0.988	1163	0.988
258	0.972	409	0.988	560	0.988	711	0.987	862	0.987	1013	0.988	1164	0.988
259	0.972	410	0.988	561	0.988	712	0.987	863	0.987	1014	0.988	1165	0.988
260	0.972	411	0.988	562	0.988	713	0.988	864	0.987	1015	0.987	1166	0.988
261	0.972	412	0.988	563	0.988	714	0.988	865	0.987	1016	0.987	1167	0.987
262	0.971	413	0.988	564	0.988	715	0.988	866	0.987	1017	0.987	1168	0.987
263	0.971	414	0.988	565	0.988	716	0.988	867	0.987	1018	0.987	1169	0.987
264	0.971	415	0.988	566	0.988	717	0.988	868	0.987	1019	0.987	1170	0.987
265	0.971	416	0.988	567	0.988	718	0.988	869	0.987	1020	0.987	1171	0.987
266	0.970	417	0.988	568	0.988	719	0.988	870	0.988	1021	0.987	1172	0.987
267	0.970	418	0.988	569	0.988	720	0.988	871	0.988	1022	0.987	1173	0.987
268	0.969	419	0.988	570	0.988	721	0.987	872	0.988	1023	0.987	1174	0.986
269	0.969	420	0.988	571	0.988	722	0.987	873	0.988	1024	0.987	1175	0.986
270	0.969	421	0.988	572	0.988	723	0.987	874	0.988	1025	0.987	1176	0.986
271	0.968	422	0.988	573	0.988	724	0.987	875	0.988	1026	0.987	1177	0.986
272	0.968	423	0.988	574	0.988	725	0.987	876	0.988	1027	0.987	1178	0.986
273	0.968	424	0.988	575	0.988	726	0.987	877	0.988	1028	0.987	1179	0.986
274	0.967	425	0.988	576	0.988	727	0.987	878	0.988	1029	0.987	1180	0.986
275	0.967	426	0.988	577	0.988	728	0.987	879	0.988	1030	0.987	1181	0.986
276	0.967	427	0.988	578	0.988	729	0.987	880	0.988	1031	0.987	1182	0.986
277	0.967	428	0.988	579	0.988	730	0.987	881	0.988	1032	0.987	1183	0.986
278	0.968	429	0.988	580	0.988	731	0.988	882	0.988	1033	0.987	1184	0.986
279	0.968	430	0.988	581	0.988	732	0.988	883	0.988	1034	0.986	1185	0.986
280	0.969	431	0.988	582	0.988	733	0.988	884	0.988	1035	0.987	1186	0.986
281	0.970	432	0.988	583	0.988	734	0.988	885	0.988	1036	0.987	1187	0.986
282	0.971	433	0.988	584	0.988	735	0.988	886	0.988	1037	0.986	1188	0.985
283	0.971	434	0.988	585	0.988	736	0.988	887	0.988	1038	0.987	1189	0.986
284	0.972	435	0.988	586	0.988	737	0.988	888	0.988	1039	0.987	1190	0.986
285	0.972	436	0.988	587	0.988	738	0.988	889	0.988	1040	0.987	1191	0.986
286	0.972	437	0.988	588	0.988	739	0.988	890	0.988	1041	0.987	1192	0.986
287	0.973	438	0.988	589	0.988	740	0.988	891	0.987	1042	0.987	1193	0.986
288	0.973	439	0.988	590	0.988	741	0.988	892	0.987	1043	0.987	1194	0.986
289	0.973	440	0.988	591	0.988	742	0.988	893	0.987	1044	0.987	1195	0.986
290	0.974	441	0.988	592	0.988	743	0.988	894	0.986	1045	0.987	1196	0.986
291	0.974	442	0.988	593	0.988	744	0.987	895	0.986	1046	0.987	1197	0.986

(Continued)

Table 8.13 (*Continued*).

nm	R	nm	R	nm	R	nm	R	nm	R	nm	R	nm	R
292	0.975	443	0.988	594	0.988	745	0.987	896	0.986	1047	0.986	1198	0.986
293	0.975	444	0.988	595	0.988	746	0.987	897	0.985	1048	0.986	1199	0.986
294	0.976	445	0.988	596	0.988	747	0.987	898	0.985	1049	0.986	1200	0.986
295	0.976	446	0.988	597	0.988	748	0.988	899	0.985	1050	0.986	1201	0.986
296	0.976	447	0.988	598	0.988	749	0.988	900	0.985	1051	0.987	1202	0.986
297	0.977	448	0.988	599	0.988	750	0.988	901	0.986	1052	0.987	1203	0.986
298	0.977	449	0.988	600	0.988	751	0.988	902	0.986	1053	0.987	1204	0.986
299	0.977	450	0.988	601	0.988	752	0.988	903	0.986	1054	0.988	1205	0.986
300	0.977	451	0.988	602	0.988	753	0.988	904	0.987	1055	0.988	1206	0.985
301	0.978	452	0.988	603	0.988	754	0.988	905	0.987	1056	0.988	1207	0.985
302	0.978	453	0.988	604	0.988	755	0.988	906	0.987	1057	0.989	1208	0.984
303	0.978	454	0.988	605	0.988	756	0.988	907	0.987	1058	0.989	1209	0.984
304	0.978	455	0.988	606	0.988	757	0.988	908	0.987	1059	0.989	1210	0.984
305	0.979	456	0.988	607	0.988	758	0.988	909	0.987	1060	0.989	1211	0.984
306	0.979	457	0.988	608	0.988	759	0.988	910	0.987	1061	0.988	1212	0.984
307	0.979	458	0.988	609	0.988	760	0.988	911	0.986	1062	0.988	1213	0.985
308	0.979	459	0.988	610	0.988	761	0.987	912	0.986	1063	0.988	1214	0.985
309	0.980	460	0.988	611	0.988	762	0.987	913	0.986	1064	0.987	1215	0.986
310	0.980	461	0.988	612	0.988	763	0.987	914	0.986	1065	0.987	1216	0.986
311	0.980	462	0.988	613	0.988	764	0.988	915	0.986	1066	0.987	1217	0.986
312	0.981	463	0.988	614	0.988	765	0.988	916	0.986	1067	0.986	1218	0.986
313	0.981	464	0.988	615	0.988	766	0.988	917	0.987	1068	0.986	1219	0.986
314	0.981	465	0.988	616	0.988	767	0.988	918	0.987	1069	0.987	1220	0.986
315	0.982	466	0.988	617	0.988	768	0.988	919	0.987	1070	0.987	1221	0.987
316	0.982	467	0.988	618	0.988	769	0.988	920	0.987	1071	0.987	1222	0.987
317	0.983	468	0.988	619	0.988	770	0.988	921	0.987	1072	0.987	1223	0.987
318	0.983	469	0.988	620	0.988	771	0.988	922	0.987	1073	0.987	1224	0.987
319	0.983	470	0.988	621	0.988	772	0.988	923	0.987	1074	0.987	1225	0.987
320	0.984	471	0.988	622	0.988	773	0.988	924	0.987	1075	0.987	1226	0.987
321	0.984	472	0.988	623	0.988	774	0.988	925	0.987	1076	0.987	1227	0.987
322	0.984	473	0.988	624	0.988	775	0.988	926	0.987	1077	0.987	1228	0.987
323	0.984	474	0.988	625	0.988	776	0.988	927	0.987	1078	0.987	1229	0.987
324	0.984	475	0.988	626	0.988	777	0.988	928	0.987	1079	0.987	1230	0.987
325	0.985	476	0.988	627	0.988	778	0.988	929	0.988	1080	0.987	1231	0.986
326	0.985	477	0.989	628	0.988	779	0.988	930	0.988	1081	0.987	1232	0.986
327	0.985	478	0.989	629	0.988	780	0.988	931	0.988	1082	0.987	1233	0.985
328	0.985	479	0.989	630	0.988	781	0.988	932	0.988	1083	0.987	1234	0.985
329	0.985	480	0.989	631	0.988	782	0.988	933	0.988	1084	0.987	1235	0.985
330	0.985	481	0.989	632	0.988	783	0.988	934	0.988	1085	0.987	1236	0.985
331	0.985	482	0.988	633	0.988	784	0.988	935	0.988	1086	0.987	1237	0.986
332	0.985	483	0.988	634	0.988	785	0.988	936	0.987	1087	0.986	1238	0.986
333	0.985	484	0.988	635	0.988	786	0.988	937	0.987	1088	0.986	1239	0.986
334	0.985	485	0.988	636	0.988	787	0.987	938	0.987	1089	0.986	1240	0.986
335	0.986	486	0.988	637	0.988	788	0.987	939	0.987	1090	0.986	1241	0.986
336	0.986	487	0.989	638	0.988	789	0.987	940	0.987	1091	0.987	1242	0.986

(*Continued*)

Table 8.13 (*Continued*).

nm	R	nm	R	nm	R	nm	R	nm	R	nm	R	nm	R
337	0.986	488	0.989	639	0.988	790	0.987	941	0.987	1092	0.987	1243	0.986
338	0.986	489	0.989	640	0.988	791	0.987	942	0.987	1093	0.987	1244	0.986
339	0.986	490	0.989	641	0.988	792	0.987	943	0.987	1094	0.988	1245	0.986
340	0.986	491	0.989	642	0.988	793	0.988	944	0.987	1095	0.988	1246	0.986
341	0.986	492	0.989	643	0.988	794	0.988	945	0.988	1096	0.989	1247	0.986
342	0.986	493	0.989	644	0.988	795	0.988	946	0.988	1097	0.989	1248	0.986
343	0.986	494	0.989	645	0.988	796	0.988	947	0.988	1098	0.989	1249	0.985
344	0.986	495	0.989	646	0.988	797	0.988	948	0.988	1099	0.989	1250	0.985
345	0.986	496	0.988	647	0.988	798	0.988	949	0.988	1100	0.989	1251	0.985
346	0.987	497	0.988	648	0.988	799	0.988	950	0.988	1101	0.989	1252	0.985
347	0.987	498	0.988	649	0.988	800	0.988	951	0.988	1102	0.988	1253	0.984
348	0.987	499	0.988	650	0.988	801	0.988	952	0.988	1103	0.988	1254	0.984
349	0.987	500	0.988	651	0.988	802	0.987	953	0.988	1104	0.987	1255	0.984
350	0.987	501	0.988	652	0.988	803	0.987	954	0.988	1105	0.987	1256	0.984
351	0.987	502	0.988	653	0.988	804	0.988	955	0.988	1106	0.987	1257	0.985
352	0.987	503	0.988	654	0.988	805	0.988	956	0.988	1107	0.988	1258	0.985
353	0.987	504	0.988	655	0.988	806	0.988	957	0.989	1108	0.988	1259	0.985
354	0.987	505	0.989	656	0.988	807	0.988	958	0.988	1109	0.988	1260	0.985
355	0.987	506	0.989	657	0.988	808	0.988	959	0.988	1110	0.989	1261	0.986
356	0.987	507	0.989	658	0.988	809	0.988	960	0.987	1111	0.989	1262	0.986
357	0.987	508	0.989	659	0.988	810	0.988	961	0.987	1112	0.989	1263	0.987
358	0.987	509	0.989	660	0.988	811	0.988	962	0.986	1113	0.988	1264	0.987
359	0.987	510	0.988	661	0.988	812	0.988	963	0.986	1114	0.988	1265	0.987
360	0.987	511	0.988	662	0.988	813	0.988	964	0.986	1115	0.988	1266	0.987
361	0.987	512	0.988	663	0.988	814	0.988	965	0.986	1116	0.987	1267	0.986
362	0.987	513	0.988	664	0.988	815	0.988	966	0.986	1117	0.987	1268	0.986
363	0.987	514	0.988	665	0.988	816	0.988	967	0.987	1118	0.987	1269	0.986
364	0.987	515	0.988	666	0.988	817	0.987	968	0.988	1119	0.987	1270	0.986
365	0.987	516	0.988	667	0.988	818	0.987	969	0.988	1120	0.987	1271	0.986
366	0.987	517	0.988	668	0.988	819	0.987	970	0.988	1121	0.987	1272	0.986
367	0.987	518	0.988	669	0.988	820	0.988	971	0.988	1122	0.987	1273	0.986
368	0.987	519	0.988	670	0.988	821	0.988	972	0.988	1123	0.987	1274	0.986
369	0.987	520	0.988	671	0.988	822	0.988	973	0.988	1124	0.988	1275	0.986
370	0.987	521	0.988	672	0.987	823	0.988	974	0.988	1125	0.988	1276	0.986
371	0.987	522	0.988	673	0.987	824	0.987	975	0.988	1126	0.988	1277	0.986
372	0.987	523	0.988	674	0.987	825	0.987	976	0.987	1127	0.988	1278	0.986
373	0.988	524	0.988	675	0.987	826	0.987	977	0.987	1128	0.988	1279	0.985
374	0.988	525	0.988	676	0.988	827	0.987	978	0.987	1129	0.988	1280	0.985
375	0.988	526	0.988	677	0.988	828	0.987	979	0.987	1130	0.988	1281	0.985
376	0.988	527	0.988	678	0.988	829	0.987	980	0.987	1131	0.988	1282	0.985
377	0.988	528	0.988	679	0.988	830	0.987	981	0.987	1132	0.988	1283	0.985
378	0.988	529	0.988	680	0.988	831	0.987	982	0.987	1133	0.988	1284	0.984
379	0.987	530	0.988	681	0.988	832	0.987	983	0.987	1134	0.988	1285	0.984
380	0.987	531	0.988	682	0.988	833	0.987	984	0.987	1135	0.988	1286	0.984
381	0.987	532	0.988	683	0.988	834	0.987	985	0.987	1136	0.988	1287	0.984

(*Continued*)

Table 8.13 (*Continued*).

nm	R	nm	R	nm	R	nm	R	nm	R	nm	R	nm	R
382	0.987	533	0.988	684	0.988	835	0.987	986	0.987	1137	0.988	1288	0.984
383	0.987	534	0.988	685	0.988	836	0.987	987	0.987	1138	0.988	1289	0.984
384	0.987	535	0.988	686	0.988	837	0.987	988	0.988	1139	0.988	1290	0.985
385	0.988	536	0.988	687	0.988	838	0.987	989	0.987	1140	0.988	1291	0.985
386	0.988	537	0.988	688	0.988	839	0.987	990	0.987	1141	0.987	1292	0.985
387	0.988	538	0.988	689	0.988	840	0.987	991	0.987	1142	0.987	1293	0.986
388	0.988	539	0.988	690	0.988	841	0.987	992	0.987	1143	0.987	1294	0.986
389	0.988	540	0.988	691	0.988	842	0.987	993	0.987	1144	0.986	1295	0.986
390	0.988	541	0.988	692	0.988	843	0.986	994	0.986	1145	0.986	1296	0.986
391	0.988	542	0.988	693	0.988	844	0.986	995	0.986	1146	0.986	1297	0.986
392	0.988	543	0.988	694	0.987	845	0.986	996	0.986	1147	0.986	1298	0.986
393	0.988	544	0.988	695	0.987	846	0.986	997	0.986	1148	0.986	1299	0.985
394	0.988	545	0.988	696	0.987	847	0.986	998	0.986	1149	0.986	1300	0.985
395	0.988	546	0.988	697	0.988	848	0.987	999	0.985	1150	0.986		
396	0.988	547	0.988	698	0.988	849	0.987	1000	0.985	1151	0.987		
397	0.988	548	0.988	699	0.988	850	0.988	1001	0.985	1152	0.987		
398	0.988	549	0.988	700	0.988	851	0.988	1002	0.986	1153	0.987		
399	0.989	550	0.988	701	0.988	852	0.988	1003	0.986	1154	0.987		
400	0.989	551	0.988	702	0.988	853	0.988	1004	0.986	1155	0.988		

8.4 Ultraviolet Stray Light Measurement Standards

Multiple standards are used for measuring stray light when applied to ultraviolet or visible spectrophotometric measurements. The following materials are used for standards and are traceable to NIST SRM 2032. These are listed in order from lowest wavelength cut-off to highest wavelength cut-off. Sodium Chloride (NaCl), Potassium Chloride (KCl), Lithium Carbonate (Li_2CO_3), Potassium Iodide (KI), Sodium Iodide (NaI), Acetone (CH_3-C=O-CH_3), Sodium Nitrite ($NaNO_2$).

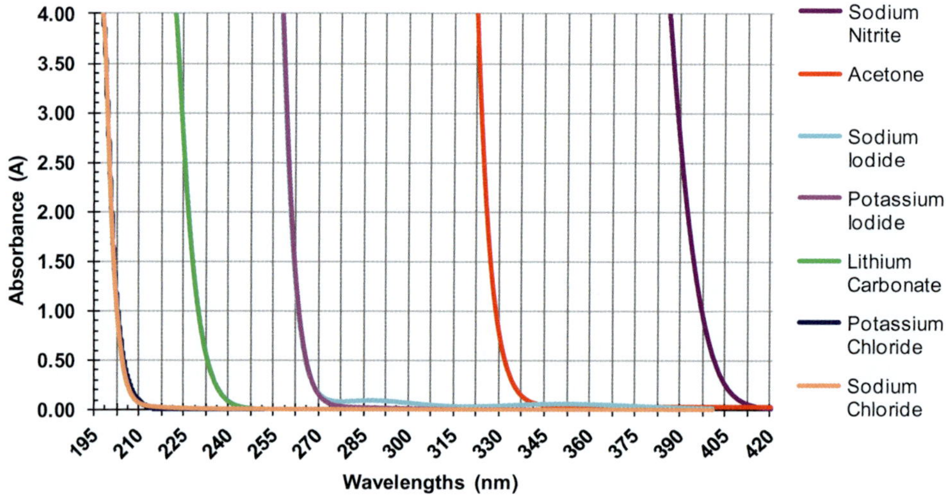

Fig. 8.12 Comparative series of stray light standards traceable to SRM 2032. (Data and spectrum provided with permission by Starna Ltd., Hainault, IG6 3UT, UK. www.starna.com).

Table 8.14 Ultraviolet stray light measurement standards: Cutoff wavelengths and composition for different materials. (Data and spectrum provided with permission by Starna Ltd., Hainault, IG6 3UT, UK. www.starna.com).

Material	Cutoff wavelength (nm)	Concentration
Sodium Nitrite	390	5% aqueous
Acetone	326	Spectral Grade
Sodium Iodide	260	1% aqueous
Potassium Iodide	260	1% aqueous
Lithium Carbonate	227	saturated
Potassium Chloride	200	1.2% aqueous
Sodium Chloride	205	1% aqueous

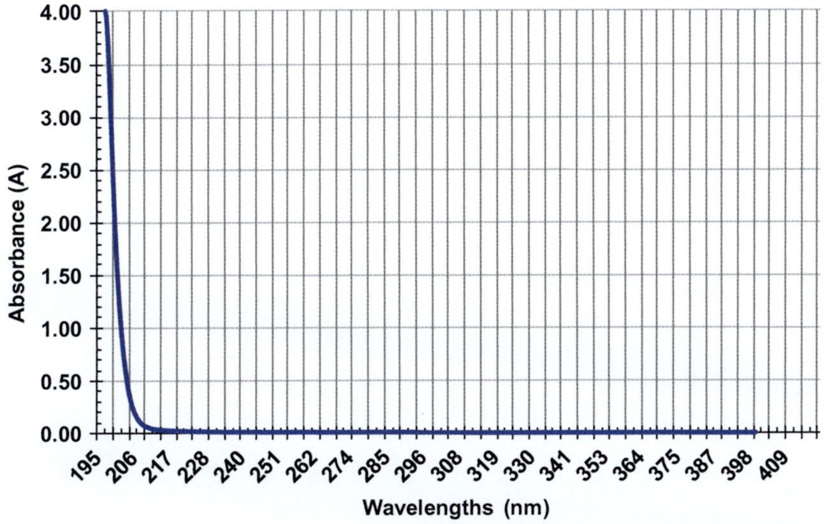

Fig. 8.13 Sodium chloride stray light standard traceable to SRM 2032 (Data and spectrum provided with permission by Starna Ltd., Hainault, IG6 3UT, UK. www.starna.com).

Table 8.15 Data table for sodium chloride stray light standard traceable to SRM 2032 (Data and spectrum provided with permission by Starna Ltd., Hainault, IG6 3UT, UK. www.starna.com).

Wavelength (nm)	Actual absorbance
204.00	0.4935
202.20	0.9960
201.10	1.4916
200.30	1.9775
199.60	2.5132
199.00	3.0264
198.40	3.5308
197.30	4.0104

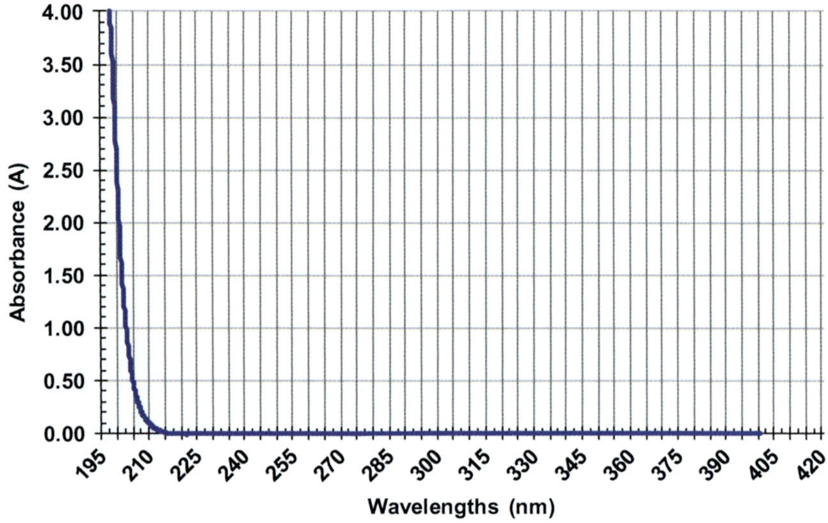

Fig. 8.14 Potassium chloride stray light standard traceable to SRM 2032 (Data and spectrum provided with permission by Starna Ltd., Hainault, IG6 3UT, UK. www.starna.com).

Table 8.16 Data table for potassium chloride stray light standard traceable to SRM 2032 (Data and spectrum provided with permission by Starna Ltd., Hainault, IG6 3UT, UK. www.starna.com).

Wavelength (nm)	Actual absorbance
204.40	0.5002
202.40	0.9857
201.10	1.5170
200.20	2.0305
199.50	2.5342
198.90	3.0162
198.30	3.5130
197.50	4.0138

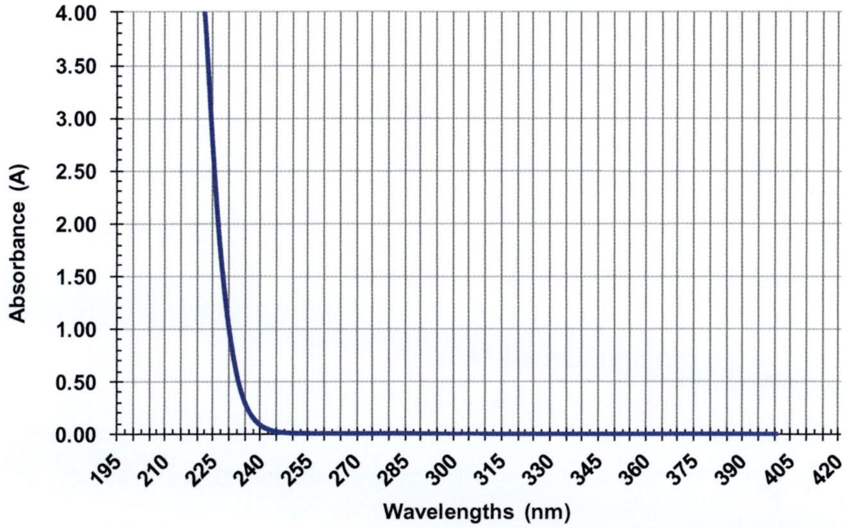

Fig. 8.15 Lithium carbonate stray light standard traceable to SRM 2032 (Data and spectrum provided with permission by Starna Ltd., Hainault, IG6 3UT, UK. www.starna.com).

Table 8.17 Data table for lithium carbonate stray light standard traceable to SRM 2032 (Data and spectrum provided with permission by Starna Ltd., Hainault, IG6 3UT, UK. www.starna.com).

Wavelength (nm)	Actual absorbance
232.60	0.5016
229.60	1.0063
227.70	1.5074
226.30	1.9929
225.10	2.5001
224.10	2.9863
223.10	3.5156
222.00	3.9934

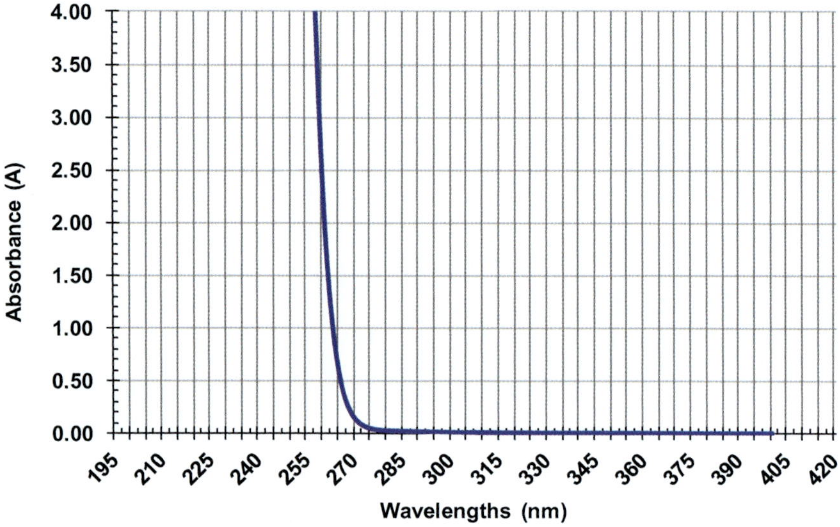

Fig. 8.16 Potassium iodide stray light standard traceable to SRM 2032 (Data and spectrum provided with permission by Starna Ltd., Hainault, IG6 3UT, UK. www.starna.com).

Table 8.18 Data table for potassium iodide stray light standard traceable to SRM 2032 (Data and spectrum provided with permission by Starna Ltd., Hainault, IG6 3UT, UK. www.starna.com).

Wavelength (nm)	Actual absorbance
265.50	0.4941
263.00	1.0026
261.50	1.5129
260.40	2.0194
259.60	2.4789
258.80	3.0273
258.20	3.4837
257.50	3.9956

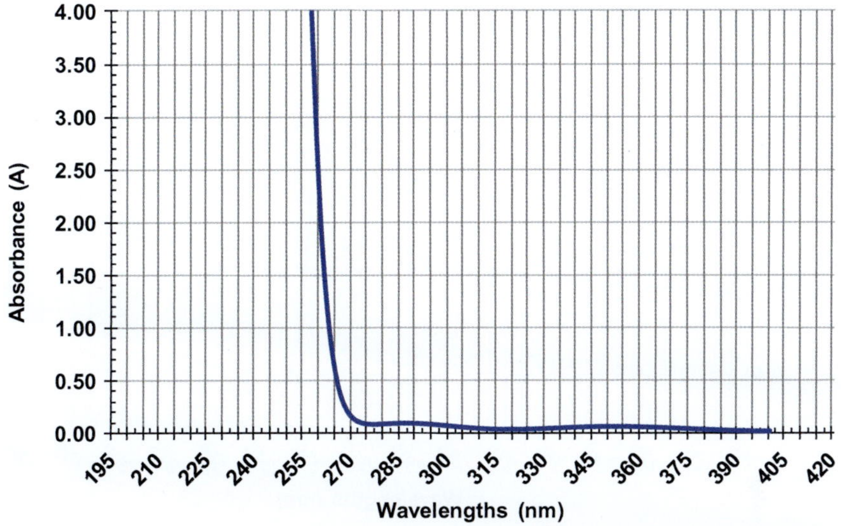

Fig. 8.17 Sodium iodide stray light standard traceable to SRM 2032 (Data and spectrum provided with permission by Starna Ltd., Hainault, IG6 3UT, UK. www.starna.com).

Table 8.19 Data table for sodium iodide stray light standard traceable to SRM 2032 (Data and spectrum provided with permission by Starna Ltd., Hainault, IG6 3UT, UK. www.starna.com).

Wavelength (nm)	Actual absorbance
265.50	0.4939
263.00	0.9931
261.50	1.4965
260.40	1.9991
259.50	2.5130
258.80	2.9947
258.10	3.5161
257.50	3.9622

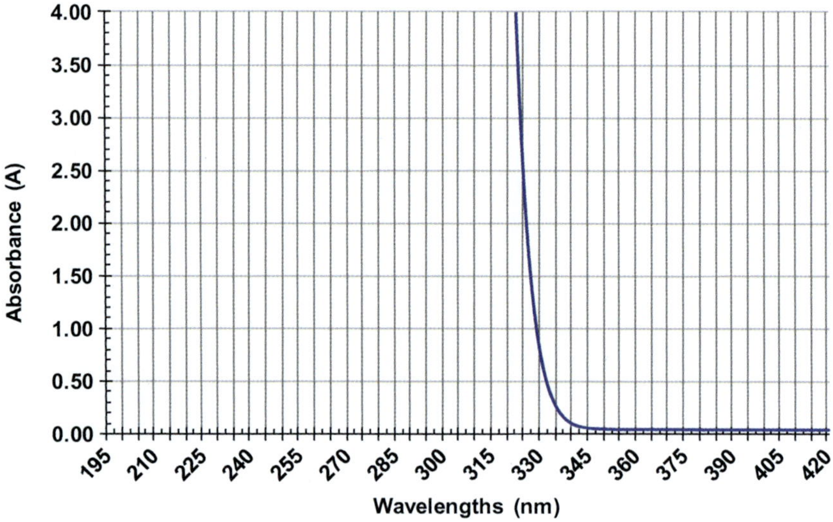

Fig. 8.18 Acetone stray light standard traceable to SRM 2032 (Data and spectrum provided with permission by Starna Ltd., Hainault, IG6 3UT, UK. www.starna.com).

Table 8.20 Data table for acetone stray light standard traceable to SRM 2032 (Data and spectrum provided with permission by Starna Ltd., Hainault, IG6 3UT, UK. www.starna.com).

Wavelength (nm)	Actual absorbance
331.60	0.5041
328.60	0.9944
326.80	1.4914
325.50	1.9903
324.40	2.5275
323.60	2.9902
322.80	3.5200
322.10	4.0037

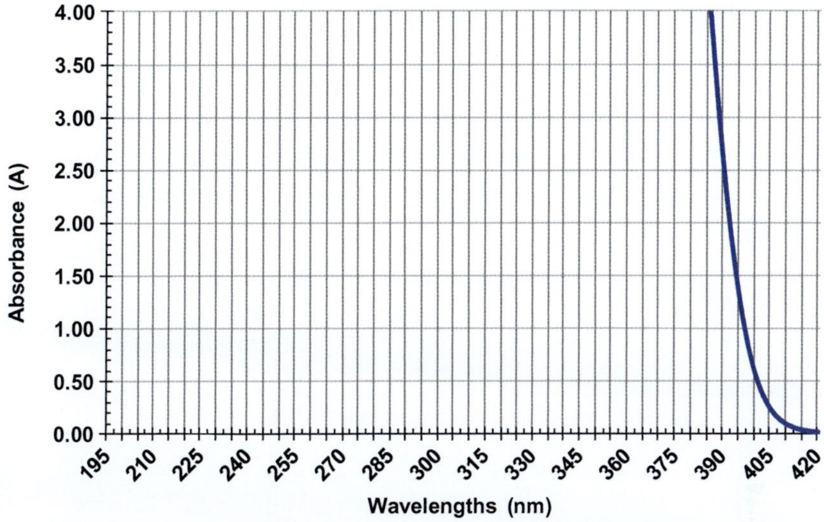

Fig. 8.19 Sodium nitrite stray light standard traceable to SRM 2032 (Data and spectrum provided with permission by Starna Ltd., Hainault, IG6 3UT, UK. www.starna.com).

Table 8.21 Data table for sodium nitrite stray light standard traceable to SRM 2032 (Data and spectrum provided with permission by Starna Ltd., Hainault, IG6 3UT, UK. www.starna.com).

Wavelength (nm)	Actual absorbance
401.10	0.4962
396.90	0.9946
394.10	1.5027
392.00	1.9947
390.20	2.5008
388.60	3.0101
387.20	3.5065
386.10	3.9805

8.5 Ultraviolet Resolution Measurement Standards

Resolution standards are used to measure or adjust the spectral bandwidth (or resolution) for a spectrophotometer. The resolution standards shown in the following figures allow the user to estimate the current resolution and adjust it as needed to configure the instrument for specific applications requiring a given resolution specification.

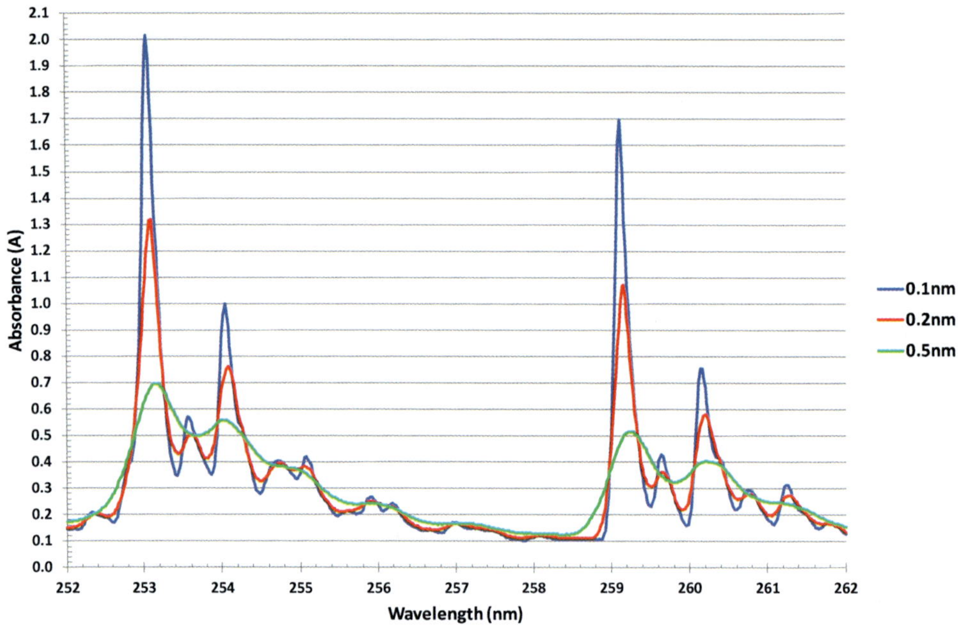

Fig. 8.20 Benzene vapor resolution standard at 0.1, 0.2, and 0.5 nm spectral bandwidth (SBW, no NIST SRM is available for comparison) (Data and spectrum provided with permission by Starna Ltd., Hainault, IG6 3UT, UK. www.starna.com).

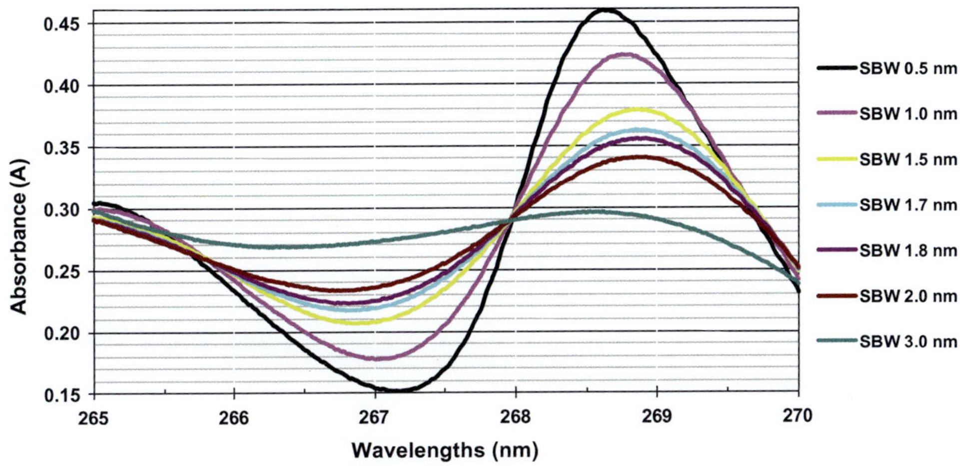

Fig. 8.21a Toluene in hexane resolution standard traceable at various SBW settings comparable to SRM 935a (Data and spectrum provided with permission by Starna Ltd., Hainault, IG6 3UT, UK. www.starna.com).

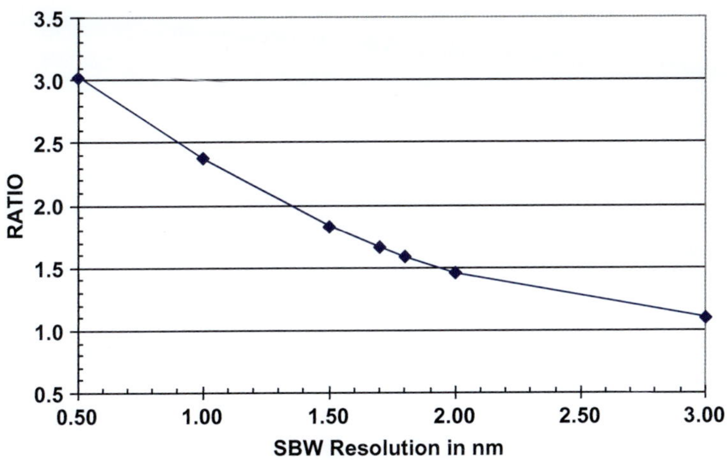

Fig. 8.21b Toluene in hexane resolution standard traceable at various SBW settings comparable to SRM 935a. The figure demonstrates the relationship between the SBW resolution (X-axis) and the Band Ratios (Y-axis). Data given in Table 8.22. (Data and spectrum provided with permission by Starna Ltd., Hainault, IG6 3UT, UK. www.starna.com).

Table 8.22 Data table for toluene in hexane resolution standard traceable to SRM 935a. Ratio of absorbances of bands in column 2 indicates useful resolution as SBW in nm. (Data and spectrum provided with permission by Starna Ltd., Hainault, IG6 3UT, UK. www.starna.com).

SBW	Bands (nm)	A	Ratio
0.5 nm	268.62	0.45954	3.02
	267.10	0.15192	
1.0 nm	268.76	0.42302	2.38
	267.02	0.17762	
1.5 nm	268.86	0.37857	1.83
	266.86	0.20662	
1.7 nm	268.86	0.36254	1.67
	266.80	0.21737	
1.8 nm	268.86	0.35505	1.59
	266.82	0.22295	
2.0 nm	268.86	0.34089	1.46
	266.76	0.23303	
3.0 nm	268.52	0.29574	1.10
	266.28	0.26831	

8.6 Fluorescent and Raman Reference Standards

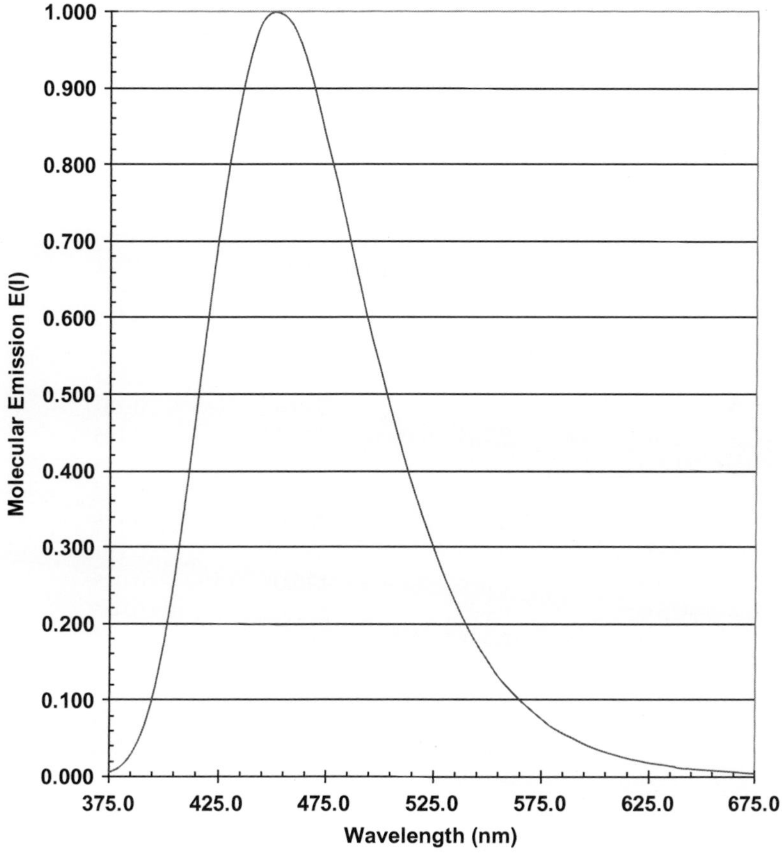

Fig. 8.22 SRM 936a certified values for quinine sulfate (sulphate) solution wavelength, intensity, and linearity standard. This is traceable to SRM 2034 and SRM 930e (Data and spectrum provided with permission by Starna Ltd., Hainault, IG6 3UT, UK. www.starna.com).

Table 8.23 Data table for quinine sulfate (sulphate) solution wavelength, intensity, and linearity standard. This reference material consists of a borosilicate glass tube, which has been permanently sealed. It is filled with a 0.5 mg/ml solution of quinine sulphate (NIST SRM 936a) in 0.105 mol/l perchloric acid. The reference material is designed for use in the evaluation of methods and the calibration of fluorescence spectrometers. The data shown below details the corrected molecular emission spectrum provided for SRM 936a. (Data and spectrum provided with permission by Starna Ltd., Hainault, IG6 3UT, UK. www.starna.com).

nm	E (l)	nm	E (l)
375.0	0.005	525.0	0.302
380.0	0.012	530.0	0.264
385.0	0.028	535.0	0.231
390.0	0.057	540.0	0.201
395.0	0.103	545.0	0.175
400.0	0.170	550.0	0.153
405.0	0.257	555.0	0.132
410.0	0.359	560.0	0.116
415.0	0.471	565.0	0.101
420.0	0.586	570.0	0.088
425.0	0.694	575.0	0.076
430.0	0.792	580.0	0.065
435.0	0.874	585.0	0.057
440.0	0.940	590.0	0.050
445.0	0.984	595.0	0.043
450.0	0.999	600.0	0.037
455.0	0.997	605.0	0.032
460.0	0.982	610.0	0.028
465.0	0.947	615.0	0.024
470.0	0.897	620.0	0.021
475.0	0.838	625.0	0.018
480.0	0.782	630.0	0.016
485.0	0.719	635.0	0.014
490.0	0.657	640.0	0.011
495.0	0.595	645.0	0.010
500.0	0.541	650.0	0.009
505.0	0.486	655.0	0.008
510.0	0.434	660.0	0.007
515.0	0.386	665.0	0.006
520.0	0.342	670.0	0.005
		675.0	0.004

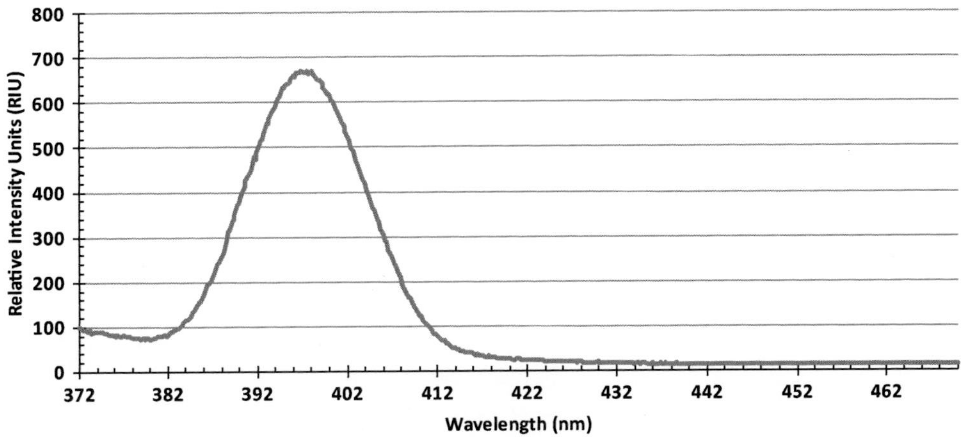

Fig. 8.23a Ultra high purity water fluorescent intensity standard. Raman water sensitivity Ex. 350 nm, Emission Peak: 396.8 nm; S/N Ratio is 1380:1 (Data and spectrum provided with permission by Starna Ltd., Hainault, IG6 3UT, UK. www.starna.com).

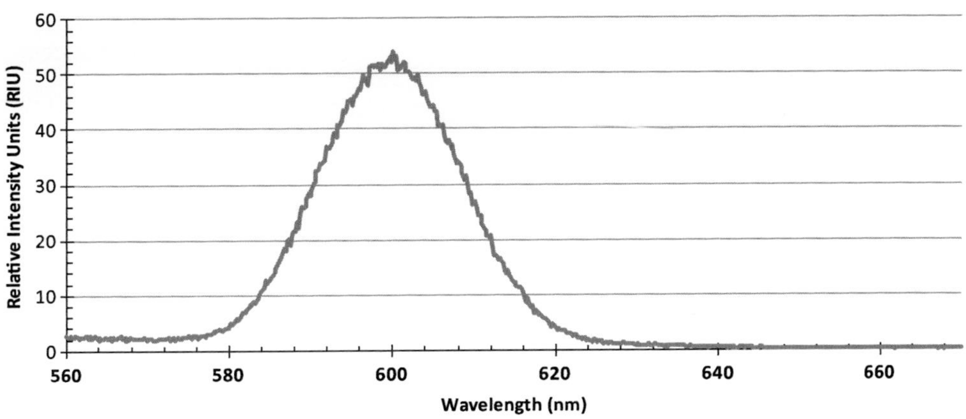

Fig. 8.23b Ultra high purity water fluorescent intensity standard. Raman water sensitivity, Excitation: 500 nm, Emission Peak: 600.1 nm; S/N Ratio is 965:1. (Data and spectrum provided with permission by Starna Ltd., Hainault, IG6 3UT, UK. www.starna.com).

Table 8.24 Data table for ultra high purity water fluorescent intensity as Raman water sensitivity. This reference material consists of a Far UV quartz cell, which has been permanently sealed. It is filled with ultra-high purity water. The reference material is designed for use in the evaluation of methods and the qualification of spectrofluorometers (fluorescence). One specific use is in the evaluation of the S/N ratio of the Raman band emission at specific wavelengths. (Data and spectrum provided with permission by Starna Ltd., Hainault, IG6 3UT, UK. www.starna.com).

Excitation Wavelength (nm)	Emission Wavelength (nm)	Signal-to-Noise Ratio	Cell Pathlength	Bandwidth	Temperature
350	396.8	1380:1	10 ± 0.01 mm	10.0 nm	23.5 ± 1°C
500	600.1	965:1	10 ± 0.01 mm	10.0 nm	23.5 ± 1°C

8.7 Optical Materials Spectra, Source Types, and Detector Specifications

The following figures and tables are included for reference. A variety of optical glasses and sapphire may be used for ultraviolet optics and sampling cells. These are presented in Tables 8.25 through 8.27; and Figures 8.24 and 8.25. Ultraviolet sources are found in Figure 8.26 and Table 8.28. Detector types are described in Figure 8.27 and Table 8.29.

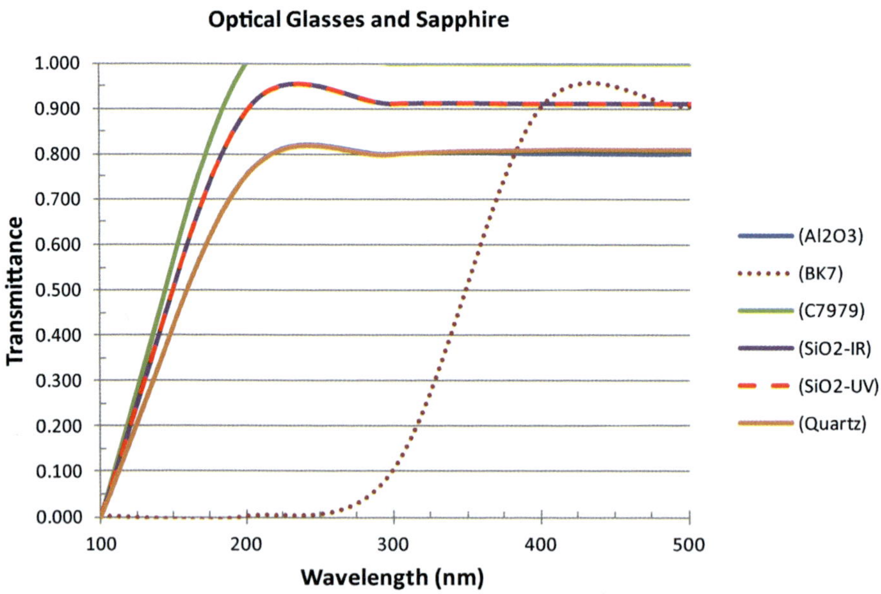

Fig. 8.24a Basic set of optical glass materials measured using transmittance geometry from 100 nm to 500 nm.

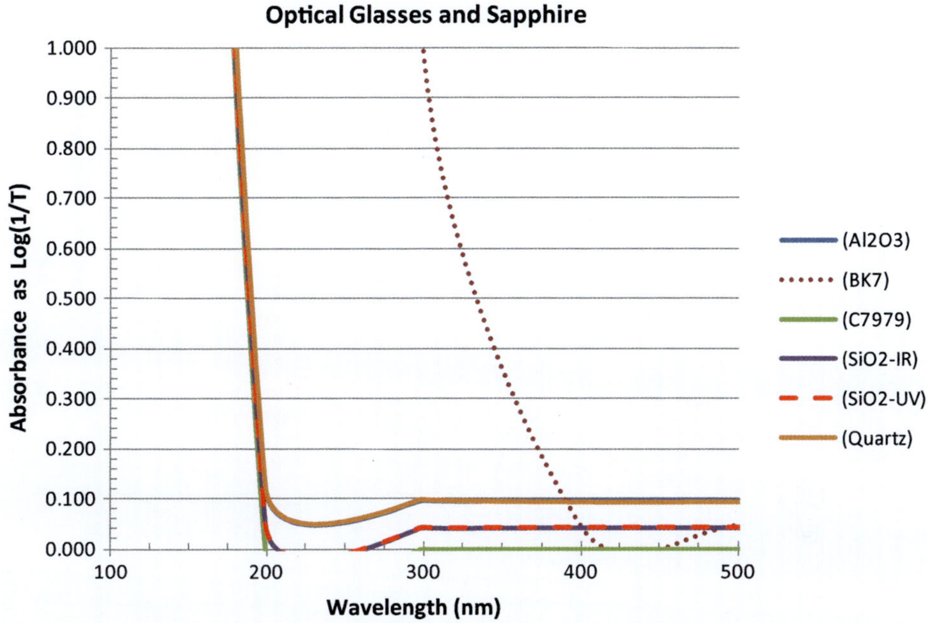

Fig. 8.24b Basic set of optical glass materials measured using absorbance as $\log_{10}$ (1/T) geometry from 100 nm to 500 nm.

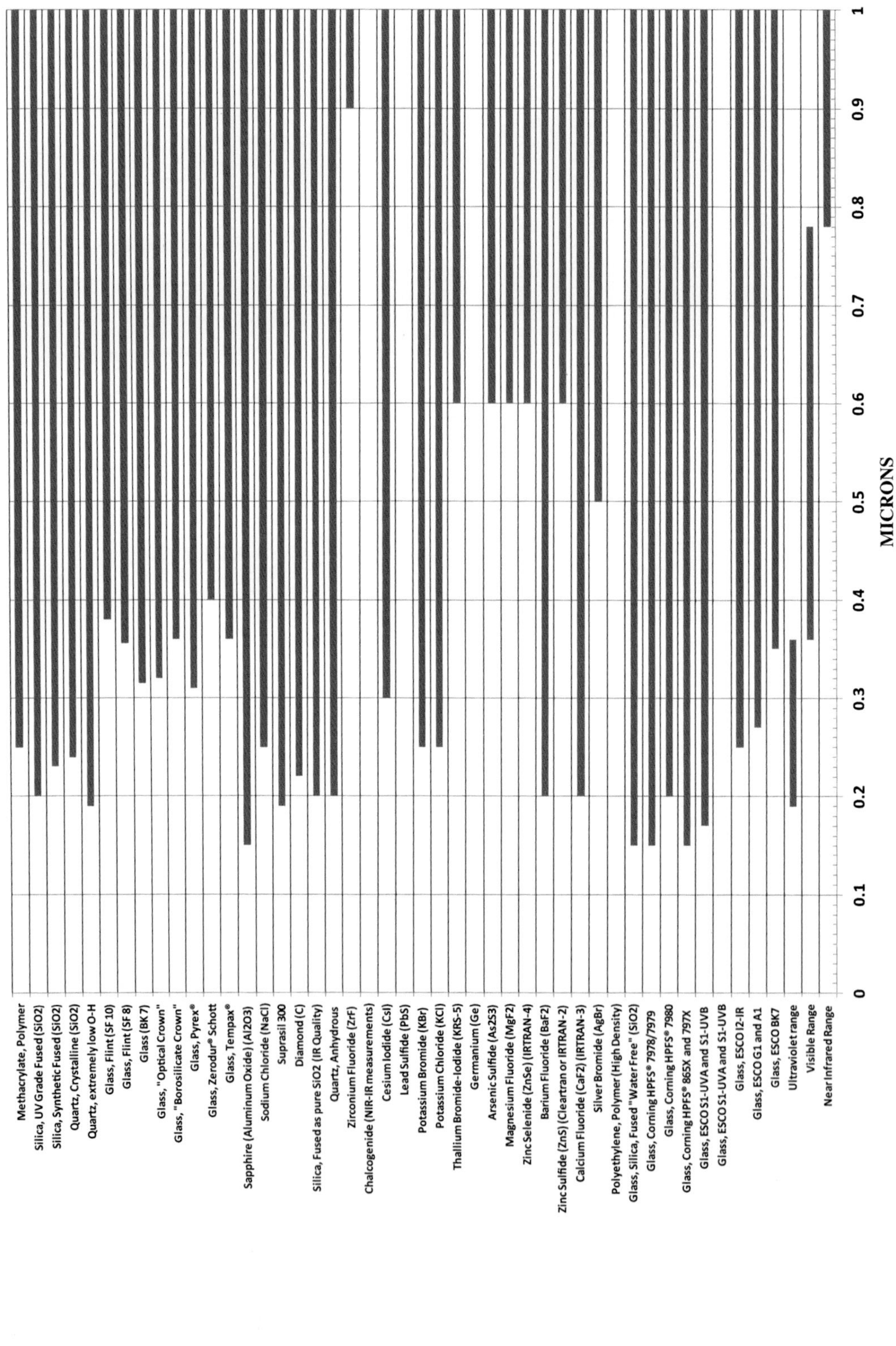

Fig. 8.25 Optical materials transmittance ranges (in microns).

Table 8.25 Optical materials transmittance ranges (in microns).

Optical material spectral ranges	Start (μm)	End (μm)	Range (μm)
Methacrylate, Polymer	0.25	1.1	1
Silica, UV Grade Fused (SiO_2)	0.2	2.5	2
Silica, Synthetic Fused (SiO_2)	0.23	2.6	2
Quartz, Crystalline (SiO_2)	0.24	2.6	2
Quartz, extremely low O-H	0.19	2.7	3
Glass, Flint (SF 10)	0.38	2.35	2
Glass, Flint (SF 8)	0.355	2.35	2
Glass (BK 7)	0.315	2.35	2
Glass, "Optical Crown"	0.32	2.3	2
Glass, "Borosilicate Crown"	0.36	2.35	2
Glass, Pyrex®	0.31	2.5	2
Glass, Tempax®	0.36	2.35	2
Sapphire (Aluminum Oxide) (Al_2O_3) 5 mm is 50 %T at 5.1 microns	0.15	5.1	5
Sodium Chloride (NaCl)	0.25	16	16
Suprasil 300	0.19	3.6	3
Diamond (C)	0.22	4	4
Silica, Fused as pure SiO_2 (IR Quality)	0.2	4	4
Quartz, Anhydrous	0.2	2.75	3
Cesium Iodide (CsI)	0.3	50	50
Potassium Bromide (KBr)	0.25	26	26
Potassium Chloride (KCl)	0.25	20	20
Barium Fluoride (BaF_2)	0.2	13	13
Calcium Fluoride (CaF_2) (IRTRAN-3)	0.2	9	9
Glass, Silica, Fused "Water Free" (SiO_2); 10 mm is 82%T at 3.15 microns	0.15	3.15	3
Glass, Corning Optical Glasses as HPFS®7978 and 7979 50%T/cm	0.15	4.35	4
Glass, Corning Optical Glasses as HPFS®7980; 50%T/cm	0.2	2.6	2
Glass, Corning Optical Glasses as HPFS®865X and 797X 50%T/cm	0.15	2.6	2
Glass, ESCO Optical Glasses as S1-UVA and S1-UVB Working Range (Band 1)	0.17	2.2	2
Glass, ESCO Optical Glasses as I2-IR Working Range	0.25	3.6	3
Glass, ESCO Optical Glasses as G1 and A1 Working Range	0.27	2.5	2
Glass, ESCO Optical Glasses as BK7 Working Range	0.35	2	2
Visible Range	0.36	0.78	0

Table 8.26 Optical materials transmittance data (in T with respect to nm).

Lambda (nm)	(Al2O3)	(BaF2)	(BK7)	(C7979)	(CaF2)	(CsI)	(CdTe)	(SiO2-IR)	(SiO2-UV)
100	0.000	0.100	0.000	0.0000	0.100	0.000	0.000	0.000	0.000
200	0.750	0.900	0.000	0.9970	0.840	0.000	0.000	0.890	0.890
300	0.800	0.900	0.100	0.9999	0.890	0.620	0.000	0.910	0.910
400	0.800	0.900	0.900	1.0000	0.900	0.740	0.000	0.910	0.910
500	0.800	0.900	0.905	1.0000	0.900	0.780	0.000	0.910	0.910
600	0.810	0.900	0.905	1.0000	0.900	0.810	0.000	0.910	0.910

Table 8.27 Optical materials transmittance data continued from Table 8.26 (in T with respect to nm).

Lambda (nm)	(KBr)	(KCl)	(KRS-5)	(KRS-6)	(LiF)	(MgF2)	(NaCl)	(Quartz)	(ZnSe)
100	0.000	0.000	0.000	0.000	0.000	0.000	0.000	0.000	0.000
200	0.650	0.650	0.000	0.000	0.900	0.780	0.100	0.750	0.000
300	0.840	0.840	0.000	0.000	0.905	0.890	0.880	0.800	0.000
400	0.895	0.895	0.000	0.000	0.905	0.920	0.900	0.810	0.000
500	0.900	0.900	0.000	0.175	0.905	0.920	0.900	0.810	0.000
600	0.900	0.900	0.000	0.280	0.905	0.920	0.900	0.810	0.650

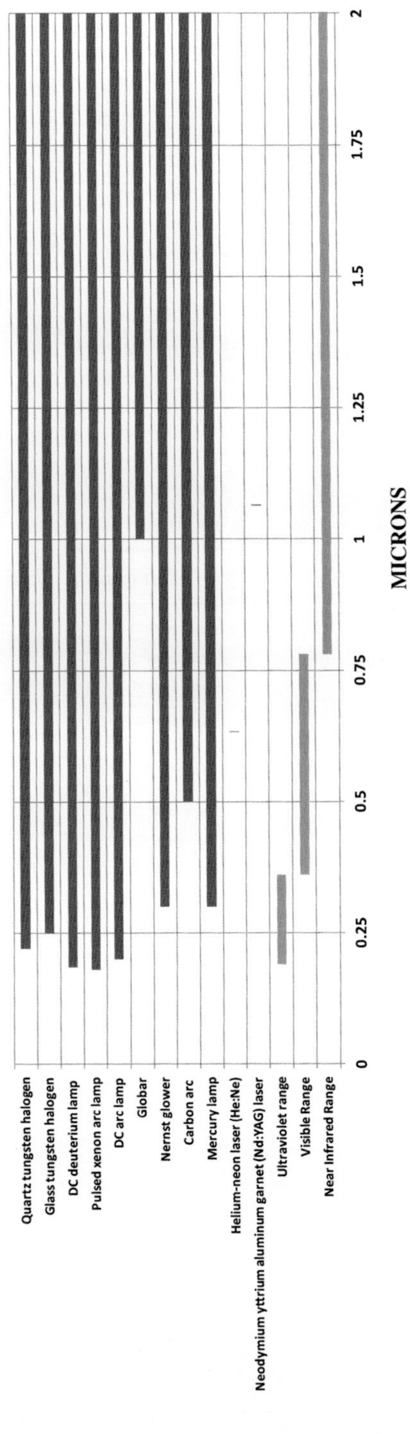

Fig. 8.26 Source emissivity ranges (in microns).

Table 8.28 Source emissivity ranges (in microns).

Source emissivity ranges	Start (μm)	End (μm)	Range (μm)
Quartz tungsten halogen	0.22	2.7	2
Glass tungsten halogen	0.25	2.25	2
DC deuterium lamp	0.185	3.75	4
Pulsed xenon arc lamp	0.18	2.5	2
DC arc lamp	0.2	2.5	2
Globar	1	100	99
Nernst glower	0.3	35	35
Carbon arc	0.5	100	100
Mercury Lamp	0.3	100	100
Helium-neon laser (He:Ne)	0.6327	0.6328	0.00010
Visible Range	0.36	0.78	0.420

Table 8.29 Detector sensitivity ranges from 0.13 microns (130 nm) to 15 microns (15000 nm).

Detector sensitivity ranges (in microns)	Start (μm)	End (μm)	Range (μm)
Silicon (Si)	0.3	1.1	0.8000
Silicon Photodiode Detectors	0.19	1.1	0.9100
Photomultipler Tube (PMT), S1 Type Photocathode	0.3	1.2	0.9000
Photomultipler Tube, Cesium Galium Arsenide (CsGaAs) Photocathode	0.3	0.93	0.6300
Photomultipler Tube, Cesium Antiminide CsSb Photocathode	0.25	0.75	0.5000
Photomultipler Tube, Multialkali (Na-K-Sb-Cs) Photocathode	0.3	0.93	0.6300
Photomultipler Tube, CsI Photocathode	0.2	0.9	0.7000
Photomultipler Tube, S1 Type Photocathode	0.3	1.2	0.9000
Photomultipler Tube, Cesium Galium Arsenide (CsGaAs) Photocathode	0.3	0.93	0.6300
Photomultipler Tube, (CsInGaAs) Photocathode	0.3	1	0.7000
Silicon (Si) CCD	0.2	1.1	0.9000
Indium Gallium Arsenide (InGaAs) CCD	0.8	1.7	0.9000
Ultraviolet range	0.19	0.36	0.1700
Visible Range	0.36	0.78	0.4200

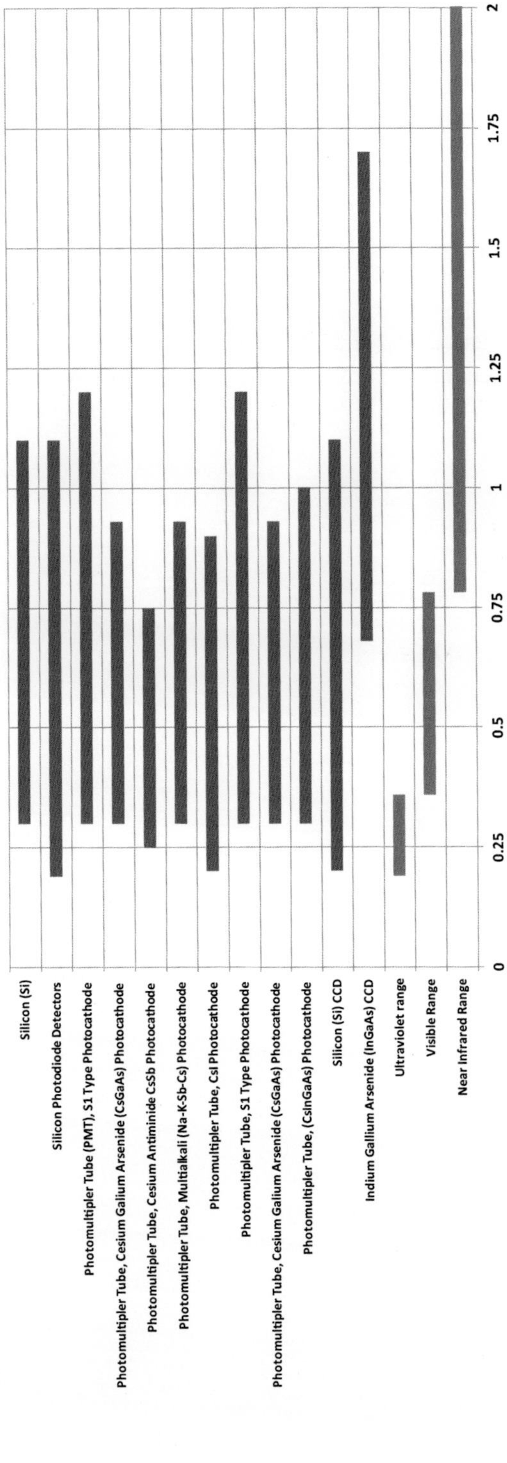

Fig. 8.27 Detector sensitivity ranges from 0 microns (0 nm) to 15 microns (15000 nm).

References

1. NIST SRM website: URL at http://www.nist.gov/srm/index3column.cfm
2. Standard Practice for Describing and Measuring Performance of Ultraviolet, Visible, and Near-Infrared Spectrophotometers. American Society for Testing and materials: ASTM International designation E 275-08 (2013), Philadelphia, PA, 2013.
3. Standard Terminology Relating to molecular Spectroscopy; American Society for Testing and materials: ASTM International designation E 131-10, 2010, Philadelphia, PA, 2010.
4. Reference spectra and data provided with permission by Starna Ltd., Hainault, IG6 3UT, UK. www.starna.com.

   Note: All figures and tables within this chapter are used with permission from Reference 4, with the exceptions of: Figures 8.24a and 8.24b are by the author; and Figure 8.25–8.27 are complied from J. Workman and A Springsteen (Eds.) Applied Spectroscopy: A compact Reference for Practitioners (Academic Press, Boston, 1998).

Chapter 9

Glossary of Terms for Ultraviolet Spectroscopy

ABSCISSA — Designates the horizontal (x) axis of a graph.

ABSORBANCE — Measure of concentration of material present: Negative log (base 10) of Transmittance [$-\log_{10}$T] of product of absorptivity (extinction coefficient), pathlength, and concentration. This is written as $A = \varepsilon bc$.
Absorbance = Log_{10}(1/Reflectance) or Absorbance = Log_{10}(1/Transmittance)

ABSORPTIVITY — Probability of light absorbing at a particular wavelength for a specific analyte under specific conditions, e.g., pH, solvent, and temperature. Thus, a specific amount of material at specified conditions will absorb a specific fraction of the light striking it. Absorptivity is generally signified by either a lower case epsilon (ε) or "a".

ACTIVE COMPONENT — In pharmaceutical preparations, the ingredient that causes the preparation to perform its desired function, (e.g., reduce fever, combat infection, etc.).

ANALYSIS — The determination of the quantity of a particular substance contained in a sample, usually expressed as weight percent. One may also analyze the optical performance of a spectrometer for its optical, mechanical, and electronic performance.

ANALYTE — The particular material (compound) or quality to be determined in an analytical analysis.

BANDPASS FILTER — A filter that is designed to provide transmission for a particular band of frequencies, but to attenuate (reduce) transmission of frequencies above (or above and below) the specified band. It has a halfwidth or half-height specified and the peak transmittance and attenuation specifications with respect to frequency or wavelength.

BANDWIDTH — This term refers to the Full-Width-at-Half-Maximum (FWHM), the linewidth, the resolution, of a spectrophotometer. It also refers to the range of frequencies (wavelengths) over which a particular instrument is designed to function within specified performance limits. Also it refers to the limits or extreme difference between the specified frequencies of a frequency range (i.e., high versus low frequencies over a range).

BEER'S LAW — Strictly defined for transmission/transmittance spectroscopy. Beer's law relates the amount of electromagnetic radiation that passes through a sample to the composition of that sample. For the case of a single absorbing species, it is often written as:

$$A = abc = \varepsilon bc \tag{9.1}$$

Where: A = measured absorbance = $-\log_{10}$ (I/Io)
I = intensity of radiation through the measuring system in the presence of the sample
Io = intensity of radiation through the measuring system in the absence of the sample
$a = \varepsilon$ = the absorptivity of the sample at the wavelength of measurement; an intrinsic property of the molecular species (also known as the extinction coefficient in physics)
b = the pathlength through the sample
c = the concentration of the sample
For a single absorbing species; if all the physical parameters (particularly the pathlength) are kept constant; then the measured absorbance (A) is directly proportional to the concentration (as weight per unit volume) of the species of interest. This is the basis of spectroscopic quantitative analysis.

CALIBRATION — Determining the response of some analytical method to known amounts of pure analyte. This term also refers to measuring and aligning the X and Y axis measurements of a spectrophotometer with respect to a reference material of standard and known characteristics, such as in wavelength or photometric accuracy calibration.

CALIBRATION CURVE — The results of a calibration when graphed, usually Cartesian coordinates, e.g., concentration (in molarity) versus absorbance. Developing a calibration curve or line often involves the application of regression methods

CONCENTRATION — The amount of analyte in a given volume of solution, e.g., moles per liter, or mass per unit volume.

CUVETTE (CUVET) — Transparent receptacle in which sample solutions are introduced into the light path of spectrometers. Usually, two sides are equal, i.e., 1 cm x 1 cm (internal dimensions), while the third dimension is elongated, possibly as long as 15 cm. For UV work, the window material is generally low OH quartz.

DETECTOR — Device used to detect the intensity of the radiation from either sample or reference beams. Usually a simple silicon diode or a more sensitive photomultiplier tube.

DOUBLE BEAM — The term applied to the simultaneous measurement of optical data from both a sample and a stable reference material. While the optical radiation used to obtain the measurements may share part or all of the optical path, there must be a separation either in space or in time between the two measurements, thus two beams of optical radiation must be used. One beam is used as a reference and the other as sample measurement. The ratio of the sample to the reference beams provide the I/Io spectral measurement for use in $-LOG_{10}(I/Io)$ computations.

ELECTROMAGNETIC SPECTRUM — The continuum of frequencies that contains electromagnetic radiation. Instruments measure the intensity of radiation within a defined range of the spectrum, and usually present the results of their measurements as a set of values of some function or the measured intensity at (usually) evenly spaced intervals within the range. The energy throughout the electromagnetic spectrum is generally from gamma radiation (most energetic and highest frequency) to radio waves (least energetic and lowest frequency).

EXTINCTION/EXTINCTION COEFFICIENT — Words sometimes substituted for absorbance and absorptivity, respectively.

FILTER SPECTROPHOTOMETER — A spectrophotometer that uses filters to isolate narrow bands of the spectrum.

FREQUENCY — The number of times per unit that the magnitude of an electromagnetic wave goes from maximum to minimum then back to maximum amplitude. The unit for the number of waves per second is Hertz (Hz).

GRATING — A reflective surface covered with evenly spaced, microscopic grooves, whose purpose is to separate individual wavelengths from broadband energy. The distance between grooves and the angle of the faces are determined by the wavelengths to be separated. The grating (except for diode arrays) is rotated at a set angle and speed and the desired wavelength is emitted through an exit slit onto the sample and detector(s).

HOLOGRAPHIC GRATING — A grating made by using a holographic process where a photoresist is placed on an optical surface such as a glass blank. A series of interference fringes corresponding to the grooves of the desired grating are recorded on this photosensitive layer by projecting a hologram of light and dark lines. Subsequent chemical treatments are used to erode either the exposed or not exposed photoresist material (there are different types such as positive and negative photoresist processes) on the surface in order to produce a grooved diffraction grating of specified lines per millimeter.

LIMIT OF DETECTION — Lowest amount of analyte that may be seen above noise level of instrument; generally estimated at three times the RMS noise level.

LINEAR — Straight line. In context, this means that for a doubling of the analyte concentration, the signal will double. This allows accurate prediction of unknown concentration from calibration curve.

LINEARITY — An experiment demonstrating that the response of an instrument changes in a predictable fashion with increasing analyte concentration.

MICRON — A unit of length in the metric system equal to 10^{-6} meter. The term micrometer is used in place of micron, but wavelength is most often expressed as in units of nanometers (10^{-9} meter) for ultraviolet through near infrared spectroscopy. The use of Angstrom (Å) as 10^{-10} meter is not common in modern spectroscopy.

NANOMETER — Unit of measurement for UV, abbreviated nm: one billionth of a meter (10^{-9} meter).

NOISE — The random component of a set of measurements, caused by the uncontrollable phenomena that affect all parts of the experimental setup. Random noise is reduced by the square root of n, where n is the number of repetitive measurements.

NORMALIZATION — This term has come to acquire two distinct meanings in spectroscopic methodology. The first meaning refers to division of each spectrum by its standard deviation so that each spectrum is given the same weight when used for calibration. The second meaning refers to the use of spectra for Discriminant Analysis, and is the method used to make the multidimensional size of the data from each material the same.

ORDER SORTING FILTER — Gratings create a spectrum more than once, and those spectra beyond the first (2nd order and up) overlap one another. Order-sorting filters remove the wavelengths above and below the desired wavelength. They are also termed high pass and low pass filters as the designation for which wavelength region they are transparent to.

PATHLENGTH — The distance the light passes through the sample in its holder. In practical terms, the inside dimension of the cuvette or sample holder. For transflectance the optical pathlength is from 2.5 to 3 times the mechanical pathlength.

RANGE — The difference between the highest and lowest values of a constituent in a sample set.

REFERENCE — In context, everything that is in the sample light path except the analyte of interest, i.e., the cuvette, solvent, and any buffer or matrix used to prepare the sample.

REFERENCE STANDARD — Analyte wherein the purity is documented and used to construct a calibration curve.

SCAN — The process of measuring a spectrum at uniform intervals throughout the spectral range of interest. The interval should be small compared with the width of the absorbance bands of the sample order to produce a quasi-continuous approximation of the sample's absorbance characteristic. It consists of measuring the transmittance, reflectance, absorbance, etc. with respect to wavelength for any material.

SCANNING — The process where the optical data is measured and recorded with respect to the wavelength range of the system. The wavelength range is measured in order, usually from lowest to highest wavelength. This usually occurs when the grating is precisely rotated about its axis.

SENSITIVITY — See "Limit of Detection".

SIGNAL — The output of the detector due to its response to light emerging from the sample holder or reference cell.

SIGNAL TO NOISE RATIO — The ratio of the value of the signal (intelligence, message, or effect to be transmitted) to that of the noise (random variation or disturbance of the signal). A measure of performance for a measureable characteristic of a device which includes a level of disturbance to the desired characteristic. In spectrometers it is often measured by taking the amplitude of the signal for 100% Transmittance or 100% Reflectance and dividing it by the root mean square deviation of the repetitive measurements of that sample.

SLIT — An aperture, usually rectangular in shape with a large length to width ratio, and fixed or adjustable shape through which radiation enters or leaves an instrument. The slit aperture is usually quite small relative to the light source.

SLIT WIDTH — Size of opening of slit through which light emerges. Size depends on wavelength range, separation ability of wavelength selector (e.g., grating), and desired isolation of specific wavelength (i.e., linewidth). Often fixed or automatically programmed.

SOLVENT — Liquid used to dissolve the sample analyzed. Commonly water or methanol of high purity. Usually designated as specially purified for UV work, e.g., "Spectrophotometer Quality" or "Spectro-Quality" or "Spectro-Grade".

SOLVENT CUTOFF — The wavelength at which the solvent absorbs a significant portion of the light, causing a loss of signal and inability to perform an analysis. In other words, the solvent becomes opaque to the wavelengths being used. This is common in the ultraviolet, much more rare in the visible.

SOURCE — Also known as the "lamp" (UV-Vis-NIR) or Globar (IR) or laser (Raman). This is the origin of the light used in the spectrometer, an incandescent wire for visible light, a deuterium gas discharge lamp for ultraviolet.

SPECTROPHOTOMETER — A device used to measure and record the spectrum of a material at uniform, and usually closely spaced intervals. This type of instrument is sometimes referred to as a spectrometer, although correct terminology is to use spectrophotometer.

SPECTRUM — Series of wavelengths of radiation, belonging to a specific portion of the electromagnetic continuum, e.g., the ultraviolet spectrum.

STRAY LIGHT — Any radiation reaching the detector that is not emitted from the sample at the chosen or selected measurement wavelength.

TRANSMISSION — The property of a material that allows radiation entering one face of the sample faces to exit through the opposite face.

TRANSMITTANCE — The fraction of radiation entering one face of a sample that exits through the opposite face; synonym for transmission.

ULTRAVIOLET — The portion of the electromagnetic spectrum detectable generally from 190 to 400 nm for the majority of laboratory work; most often 190 nm to 360 nm for the majority of ultraviolet measurements.

WAVELENGTH — The distance from one crest of an electromagnetic wave to the same position on the subsequent wave. Peak-to-peak distance, usually measured in nanometers for ultraviolet spectroscopy.

Volume 1 Index

D

E

F

G

V

W

X

Y